RICHARD ASKEY
University of Wisconsin - Madison

Orthogonal Polynomials and Special Functions

SOCIETY FOR INDUSTRIAL AND APPLIED MATHEMATICS

PHILADELPHIA, PENNSYLVANIA   1975

Printed by Odyssey Press, Dover, New Hampshire, U.S.A.

Contents

This volume is dedicated to
GABOR SZEGÖ
who has shown that special functions can be beautiful and useful at the same time

Preface

The study and use of special functions is a very old branch of mathematics and many of the great mathematicians of the past have made important contributions. One need only recall Bernoulli numbers and polynomials; Euler's gamma and beta functions; the hypergeometric function of Euler and Gauss; Abel's, Jacobi's and Weierstrass' work on elliptic functions; Bessel functions; and the polynomials of Legendre, Jacobi, Laguerre, and Hermite. Most of these functions were introduced to solve specific problems and this will be the theme of these lectures. One studies special functions not for their own sake, but to be able to use them to solve problems.

About ten years ago a few basic problems arose in the joint work I was doing with Stephen Wainger. At first the results we needed were in the literature but after a while we ran out of known results and had to learn something about special functions. This was a very unsettling experience for there were very few places to go to really learn about special functions. At least that is what we thought. Actually there were many, but the typical American graduate education which we had did not include anything about hypergeometric functions. And hypergeometric functions are the key to this subject, as I have found out after many years of fighting them. Like any part of mathematics which is very important there are many ways to look at hypergeometric functions. One which has become fashionable recently is to study them by means of algebraic methods, either Lie algebras or through Lie groups and symmetric spaces. These methods are quite powerful as shown by the new formulas which have been obtained by them. However, they are just a couple of many ways of treating these functions. Rather than try to give a survey of these methods (which I could not do very well) I will try to show how to use special functions to solve problems which can be understood by all mathematicians. They will be used to prove that $[(1 - r)(1 - s) + (1 - r)(1 - t) + (1 - s)(1 - t)]^{-\alpha-1}$ has nonnegative power series coefficients when $\alpha \geqq -\frac{1}{2}$. They will also be used to prove the inequality

$$\sum_{k=1}^{n} \frac{\sin k\theta}{k} > 0, \qquad 0 < \theta < \pi,$$

and to suggest the stronger inequality

$$\frac{d}{d\theta} \sum_{k=1}^{n} \frac{\sin k\theta}{k \sin \frac{1}{2}\theta} < 0, \qquad 0 < \theta < \pi.$$

There will be problems suggested by the isometric embedding of projective spaces in other projective spaces, by the desire to construct large classes of univalent functions, by applications to quadrature problems, and theorems on the location of zeros of trigonometric polynomials. There are also applications to combinatorial problems, statistics, physical problems, and many other areas which will only be hinted at. In short, almost every mathematician or person who uses much mathematics will be able to find something here of interest. That is the main reason for these lectures.

There are now a relatively large number of people who know a fair amount about special functions. These people can be very useful to other mathematicians and scientists. Most scientists are aware of this, but they are unaware that there could be a mathematician at their university or laboratory who knows something which they could use. And most mathematicians are totally unaware of the power of special functions. They react to a paper which contains a Bessel function or Legendre polynomial by turning immediately to the next paper. The best illustrations of this that I know are two papers of G. Szegö from the 1930's. In the first he used Bessel functions to give a very elegant proof of the conjecture of Friedrichs and Lewy that $[(1 - r)(1 - s) + (1 - r)(1 - t) + (1 - s)(1 - t)]^{-1}$ is absolutely monotonic. He proved this in 1933 and I know of no references to this paper in the literature after 1933 until 1972, when Gasper and I gave a new proof and pointed out how surprising the result was. This will be treated in Lecture 6. The second is more surprising for in 1936 Szegö solved a problem which was "open" from 1946 to 1966, when it was solved again. The problem is to find an asymptotic expression for the smallest eigenvalue of the truncated Hilbert matrix $[(i + j + 1)^{-1}]_0^N$. Szegö's paper was published in the Transactions of the American Mathematical Society, 40 (1936), pp. 450–461, at a time when there were fewer than ten American mathematical periodicals so it would have been hard to overlook. However, his proof used an asymptotic formula for Legendre polynomials off $[-1, 1]$ in an essential way. It seems that those people who would want this result ten to thirty years later did not read his paper. This problem was solved again in 1966 by H. Widom and H. Wilf, Proceedings of the American Mathematical Society, 17 (1966), pp. 338–344; errata 19 (1968), p. 1508, and their proof also used the same asymptotic formula for Legendre polynomials. They cannot be faulted for not having found Szegö's paper, for it is impossible to search the complete back literature; but it would have been nice if someone had read and remembered Szegö's paper who later became interested in inverting large matrices. I came across both of these papers in an accidental way. Szegö told me about the first in August, 1969, in Budapest and I did not look it up then. The mathematics library in Lund contains M. Riesz's reprint collection. I was there on a visit in November, 1969, and since Szegö had seemed pleased with this first paper I decided to look at it and see what it contained. Until I got to the last section before the appendix I thought it was a pretty paper but one which I would probably forget. Then in the last section the main result was translated into another question which solved a problem I had worked on for a couple of years. And the solution went in the opposite way than I thought it would. This impelled me to look carefully not only at this paper, but also at the remaining reprints of Szegö's work in Riesz's collection, and one of them was the second paper referred to above. Hopefully these lectures will show a few people how useful hypergeometric functions can be. Very few facts about hypergeometric functions are known, but these few facts can be very useful in many different contexts. So my advice is to learn something about hypergeometric functions; or, if this seems too hard or dull a task, get to know someone who knows something about them. And if you already know something about these functions, share your knowledge with a colleague or two, or a group of students. Every large university and research laboratory should have a person who not only can find things in the Bateman Project, but can fill in a few holes in this set of books. Such people can be very useful to scientists outside of mathematics. For example, in the last month I have been consulted by a chemist, a geneticist, and a statistician in the sociology department. All of these had interesting mathematical questions about special functions. Unfortunately I was only helpful to one of the three.

In any case I hope my point has been made; special functions are useful and those who need them and those who know them should start to talk to each other. These lectures primarily show how special functions can be used on various mathematical problems. There are two reasons for this. One, these are the most interesting recent applications and new results that I know. And second, the mathematical community at large needs the education on the usefulness of special functions more than most other people who could use them.

It is a pleasure to thank the many people who have helped to bring about this series of lectures. I. I. Hirschman, Jr. introduced me to ultraspherical polynomials when I was an undergraduate, and has provided encouragement through the years. S. Bochner was the ideal Ph.D. supervisor; he left me alone to write a thesis and gave me two problems to work on after finishing which were much harder than my thesis work. Stephen Wainger helped me solve these problems. He may have been able to solve them without me; I could not have solved them without his aid. The National Science Foundation, Office of Naval Research, Guggenheim Foundation, the Mathematics Research Center, Madison, the Mathematisch Centrum, Amsterdam, and the Research Committee of the Graduate School of the University of Wisconsin have provided financial aid at various times. My debt to the books of Szegö and Zygmund and the Bateman Project can only be repaid by passing on new information to the next generation. And I am fortunate in being able to thank two younger mathematicians, George Gasper and Tom Koornwinder, for helping to educate me. I introduced them to the problems which will be discussed, but by now I have learned more from them than they have from me. Gasper has also read a preliminary version of these lectures and suggested many improvements. Jim Cochran did a superb job of organizing the National Science Foundation Regional Conference at Virginia Polytechnic Institute at which these lectures were given in June 1974. And Sally Ross did an exceptional job of typing this manuscript. Many others have contributed in small ways and their help is appreciated, though they remain anonymous.

RICHARD ASKEY

LECTURE 1

Classical Results for Trigonometric Polynomials and Fourier Series and Other Isolated Results

The modern era of Fourier series started with Fejér's theorem that the Fourier series of a continuous function is uniformly Cesàro summable to the function. The basic fact behind his proof was the observation that

$$(1.1) \quad \sum_{k=0}^{n} \sin(k + \tfrac{1}{2})\theta = \frac{1 - \cos(n+1)\theta}{2\sin\tfrac{1}{2}\theta} = \frac{\sin^2\tfrac{1}{2}(n+1)\theta}{\sin\tfrac{1}{2}\theta} \geqq 0, \quad 0 \leqq \theta \leqq 2\pi.$$

Fejér used this to obtain

$$(1.2) \quad 1 + 2\sum_{k=1}^{n}\left(1 - \frac{k}{n+1}\right)\cos k\theta = \frac{1}{(n+1)}\left\{\frac{\sin\tfrac{1}{2}(n+1)\theta}{\sin\tfrac{1}{2}\theta}\right\}^2 \geqq 0.$$

Two important positive sums from the nineteenth century are Poisson's sum

$$(1.3) \quad 1 + 2\sum_{n=1}^{\infty} r^n \cos n\theta = \frac{1 - r^2}{1 - 2r\cos\theta + r^2} > 0, \quad -1 < r < 1,$$

and Jacobi's sum

$$(1.4) \quad 1 + 2\sum_{n=1}^{\infty} e^{-n^2 t}\cos n\theta = \sum_{k=1}^{\infty}(1 - e^{-2kt})(1 + 2e^{-(2k-1)t}\cos\theta + e^{-2(2k-1)t}),$$

which is positive for $t > 0$.

The last of the important positive approximate identities for Fourier series that we shall consider is de la Vallée Poussin's sum

$$(1.5) \quad 1 + 2\sum_{k=1}^{n}\frac{(n!)^2}{(n-k)!(n+k)!}\cos k\theta = \frac{(n!)^2}{(2n)!}\left(2\cos\frac{\theta}{2}\right)^{2n}.$$

This was discovered in 1908, de la Vallée Poussin [1]. The date is interesting, for we will see that H. Bateman [1] had a much more general sum in 1905, but it was not used until recently and so was not contained in the standard handbooks of formulas. More of this in Lectures 2 and 4.

Since Fejér's sum (1.1) was very useful there has been interest in finding other nonnegative trigonometric polynomials and finding uses for them. In 1910 Fejér conjectured that

(1.6)
$$\sum_{k=0}^{n} \frac{\sin (k + 1)\theta}{k + 1} > 0, \qquad\qquad 0 < \theta < \pi.$$

These sums are the partial sums of the Fourier series

(1.7)
$$\frac{\pi - \theta}{2} = \sum_{k=0}^{\infty} \frac{\sin (k + 1)\theta}{k + 1}, \qquad\qquad 0 < \theta \leqq \pi,$$

which was studied extensively because it illustrates the Gibbs' phenomenon. It is likely that graphs of the partial sums suggested this conjecture to Fejér. Proofs of (1.6) were given by Jackson [1] and Gronwall [1] and they were surprisingly hard. For some reason, and I suspect it was partly the degree of difficulty of these first proofs, many people have found new proofs of (1.6). One of the most incisive was Turán's proof of (1.6) from (1.1).

THEOREM 1.1. *If* $\sum_{k=0}^{\infty} |a_k|$ *converges and*

(1.8)
$$\sum_{k=0}^{\infty} a_k \sin (k + \tfrac{1}{2})\varphi \geqq 0, \qquad\qquad 0 \leqq \varphi \leqq \pi,$$

then

(1.9)
$$\sum_{k=0}^{\infty} a_k \frac{\sin (k + 1)\theta}{k + 1} > 0, \qquad\qquad 0 < \theta < \pi,$$

unless $a_k \equiv 0, k = 0, 1, \cdots$.

When $a_k = 1, k = 0, 1, \cdots, n, a_k = 0, k > n$, then (1.8) reduces to (1.1) and (1.9) is (1.6).

Turán's original proof [1] used complex variables; a second proof in which the kernel taking (1.8) to (1.9) was obtained and proved to be positive by estimating logarithms was given by Hyltén-Cavallius [1]. A third proof was given by Askey, Fitch and Gasper [1] and since the method of proof will be used later it will be given here.

A simple calculation shows that

(1.10)
$$\frac{d}{dx} \frac{\sin \alpha x}{\alpha (\sin x)^{\alpha}} = -\frac{\sin (\alpha - 1)x}{(\sin x)^{\alpha + 1}}.$$

Let $\alpha = 2n + 2$, $x = \tfrac{1}{2}\theta$ and integrate (1.10) to get

$$\frac{\sin (n + 1)\theta}{2(n + 1)(\sin \tfrac{1}{2}\theta)^{2n+2}} = \int_{\theta/2}^{\pi/2} \frac{\sin (2n + 1)\varphi}{(\sin \varphi)^{2n+3}} d\varphi.$$

Multiply by $2(\sin \tfrac{1}{2}\theta)^{2n+2}$ and a_n and sum to obtain

(1.11)
$$\sum_{n=0}^{\infty} \frac{a_n \sin (n + 1)\theta}{n + 1} = 2 \int_{\theta/2}^{\pi/2} \sum_{n=0}^{\infty} \left(\frac{\sin \tfrac{1}{2}\theta}{\sin \varphi} \right)^{2n+2} a_n \frac{\sin (n + \tfrac{1}{2})(2\varphi)}{\sin \varphi} d\varphi.$$

The left-hand side of (1.11) is the sum we want while the sum under the integral on the right-hand side is

$$(1.12) \qquad \sum_{n=0}^{\infty} a_n r^n \sin(n + \tfrac{1}{2})(2\varphi), \quad 0 \leq \tfrac{1}{2}\theta \leq \varphi < \tfrac{1}{2}\pi, \quad r = \frac{\sin^2 \tfrac{1}{2}\theta}{\sin^2 \varphi} \leq 1,$$

and the strict positivity of (1.12) for $\tfrac{1}{2}\theta < \varphi < \tfrac{1}{2}\pi$ follows from the nonnegativity of (1.8) and the strict positivity of the Poisson kernel

$$(1.13) \qquad \begin{aligned} &\sum_{n=0}^{\infty} r^n \sin(n + \tfrac{1}{2})\theta \sin(n + \tfrac{1}{2})\varphi \\ &= \frac{(1 - r)\sin \tfrac{1}{2}\theta \sin \tfrac{1}{2}\varphi[(1 - r)^2 + 4r(1 - \cos \tfrac{1}{2}(\theta + \varphi)\cos \tfrac{1}{2}(\theta - \varphi))]}{[1 - 2r\cos \tfrac{1}{2}(\theta + \varphi) + r^2][1 - 2r\cos \tfrac{1}{2}(\theta - \varphi) + r^2]}, \end{aligned}$$

for $0 \leq r < 1$.

There have been very few applications of (1.6) and this is explained very nicely by Turán's theorem. For it says that each application of (1.6) can be replaced by another argument which uses (1.1). For all of that (1.6) has been a very fruitful inequality because the various extensions and new proofs of it have led to a deeper understanding of the classical orthogonal polynomials and certain questions about analytic functions (see Alexander [1] and Feldheim [1]).

The inequality (1.1) for $0 \leq \theta \leq \pi$ is equivalent to

$$(1.14) \qquad 2 \sum_{k=1}^{n} \sin k\theta + \sin(n + 1)\theta \geq 0, \qquad 0 \leq \theta \leq \pi,$$

and this is

$$(1.15) \qquad S_n(\theta) + S_{n+1}(\theta) \geq 0, \qquad 0 \leq \theta \leq \pi,$$

where

$$S_n(\theta) = \sum_{k=1}^{n} \sin k\theta.$$

Unfortunately $S_n(\theta)$ changes sign in $0 \leq \theta \leq \pi$, but (1.15) and (1.2) both suggest consideration of the inequality

$$(1.16) \qquad \sum_{k=0}^{n} (n + 1 - k)\sin(k + 1)\theta \geq 0, \qquad 0 \leq \theta \leq \pi.$$

Lukács proved (1.16) before he died of flu in 1918 and Fejér [3] published his proof in 1928. This inequality also follows from (1.1), and the easiest way to see this is to consider the generating functions

$$(1.17) \qquad \sum_{n=0}^{\infty} r^n \sum_{k=0}^{n} \sin(k + \tfrac{1}{2})\theta = \frac{(1 + r)\sin \tfrac{1}{2}\theta}{(1 - r)(1 - 2r\cos \theta + r^2)}$$

and

$$(1.18) \qquad \sum_{n=0}^{\infty} r^n \sum_{k=0}^{n} (n + 1 - k)\sin(k + 1)\theta = \frac{\sin \theta}{(1 - r)^2(1 - 2r\cos \theta + r^2)}.$$

Except for a factor of $2 \cos \frac{1}{2}\theta$ one goes from (1.17) to (1.18) by multiplication by $(1 - r^2)^{-1}$, which has nonnegative power series coefficients. In his work on univalent functions Fejér [8] needed a positive Cesàro mean of the formal series

$$\sum_{n=0}^{\infty} (n + 1) \sin (n + 1)\theta,$$

and he showed that the $(C, 3)$ means of this series are positive, i.e.,

(1.19) $$\sum_{k=0}^{n} \frac{(4)_{n-k}}{(n-k)!}(k + 1) \sin (k + 1)\theta > 0, \qquad\qquad 0 < \theta < \pi,$$

where

(1.20) $$(a)_n = a(a + 1) \cdots (a + n - 1) = \Gamma(n + a)/\Gamma(a).$$

He used it to prove that $z + \sum_{n=2}^{\infty} a_n z^n$ is univalent for $|z| < 1$ if $\lim_{n \to \infty} a_n = c \geq 0$ and $\Delta^4 a_n = a_n - 4a_{n+1} + 6a_{n+2} - 4a_{n+3} + a_{n+4} \geq 0$, $n = 0, 1, \cdots$. Since series with $\sin (n + \frac{1}{2})\theta$ tend to behave more nicely than those with $\sin (n + 1)\theta$ (compare (1.1) with (1.6)) it is reasonable to suspect that a lower order Cesàro mean might give positivity for the formal series

$$\sum_{n=0}^{\infty} (n + \tfrac{1}{2}) \sin (n + \tfrac{1}{2})\theta,$$

and it does as Fejér [7] proved. The $(C, 2)$ means work:

(1.21) $$\sum_{k=0}^{n} \frac{(3)_{n-k}}{(n-k)!}(k + \tfrac{1}{2}) \sin (k + \tfrac{1}{2})\theta > 0, \qquad\qquad 0 < \theta < \pi.$$

Much earlier, Fejér [1] proved that

(1.22) $$\sum_{k=0}^{n} P_k(x) > 0, \qquad\qquad -1 < x \leq 1,$$

where $P_n(x)$ is the Legendre polynomial, and used it to show the positivity of the $(C, 2)$ means of the formal reproducing kernel $\sum_{n=0}^{\infty} (n + \frac{1}{2})P_n(x)$,

(1.23) $$\sum_{k=0}^{n} \frac{(3)_{n-k}}{(n-k)!}(k + \tfrac{1}{2})P_k(x) > 0, \qquad\qquad -1 < x \leq 1.$$

As Fejér remarked in [7], this positivity also follows from his inequality (1.21), and this was a crucial remark when one tries to tie all these results together.

   Both (1.19) and (1.21) are best possible but (1.19) can be improved when the range of θ is restricted. Schweitzer [1] completed earlier work of Fejér and Szegö when he proved

(1.24) $$\sum_{k=0}^{n} \frac{(3)_{n-k}}{(n-k)!}(k + 1) \sin (k + 1)\theta \geq 0, \qquad\qquad 0 \leq \theta \leq 2\pi/3,$$

and this is best possible in two ways. It fails for $\theta > 2\pi/3$ and the factor $(3)_{n-k}$ cannot be replaced by $(3 - \varepsilon)_{n-k}$ for any $\varepsilon > 0$ and have an inequality which holds for all n, and $0 \leq \theta \leq c, c > 0, c$ fixed.

Finally there were two relatively recent results of Vietoris [1]:

$$(1.25) \qquad \sum_{k=1}^{n} c_k \sin k\theta > 0, \qquad\qquad 0 < \theta < \pi,$$

and

$$(1.26) \qquad \sum_{k=0}^{n} c_k \cos k\theta > 0, \qquad\qquad 0 \leq \theta < \pi,$$

where

$$c_{2k} = c_{2k+1} = \frac{\left(\frac{1}{2}\right)_k}{k!}, \qquad\qquad k = 0, 1, \cdots.$$

In addition to a number of positive sums of orthogonal polynomials which were used to prove the positivity of certain Cotes' numbers there was one important result of Kogbetliantz which I knew but did not understand. To give it we must define the ultraspherical polynomials $C_n^\lambda(x)$. There are many ways of defining them, one of them being

$$(1.27) \qquad \frac{1 - r^2}{(1 - 2xr + r^2)^{\lambda+1}} = \sum_{n=0}^{\infty} \left(\frac{n+\lambda}{\lambda}\right) C_n^\lambda(x) r^n, \qquad\qquad \lambda \neq 0.$$

Kogbetliantz [1] proved the inequality

$$(1.28) \qquad \sum_{k=0}^{n} \frac{(2\lambda + 2)_{n-k}}{(n-k)!} \frac{(k+\lambda)C_k^\lambda(x)}{\lambda} > 0, \qquad -1 < x \leq 1, \quad \lambda > 0,$$

which is (1.19) when $\lambda = 1$, (1.23) when $\lambda = \frac{1}{2}$, and has (1.2) as a limiting result when $\lambda \to 0$. It was easy to see, as Kogbetliantz remarked, that (1.28) could not be improved in the sense that (1.28) would fail for $x = -1$ and n odd if $(2\lambda + 2)_{n-k}$ were replaced by $(a)_{n-k}$ with $a < 2\lambda + 2$. However, I did not feel that I understood this inequality for reasons which will be given in a later lecture. Also the proof was too sketchy for me to follow. I did not doubt this result, I just did not know how to prove it and had the feeling that there was much more going on here than I understood. After about ten years a little light began to dawn and another five years has cleared up the result to my satisfaction.

Just to confuse the reader a bit more, there is one more inequality which I learned only last year:

$$(1.29) \qquad \frac{\sin (n-1)\theta}{(n-1) \sin \theta} - \frac{\sin (n+1)\theta}{(n+1) \sin \theta} < \frac{4n}{n^2-1}\left[1 - \frac{\sin n\theta}{n \sin \theta}\right], \qquad 0 < \theta < \pi,$$

(see M. Robertson [1]). A slightly stronger inequality than (1.29) will come out of a synthesis of the above inequalities.

The reader is encouraged to stop at this point and see if any order suggests itself. If it does not (which is the likely state of affairs), then Lectures 8 and 9 can be consulted directly, or preferably, the detours which will be given in Lectures 2 to 7 can be followed, after which the answers in 8 and 9 should not be very surprising.

To give an indication that something new of the same type as above will be obtained we mention only the inequalities

(1.30) $$\frac{d}{d\theta} \sum_{k=0}^{n} \frac{\sin(k+1)\theta}{(k+1)\sin\frac{1}{2}\theta} < 0, \qquad\qquad 0 < \theta < \pi,$$

and

(1.31) $$\sum_{k=0}^{n} \frac{(3)_{n-k}}{(n-k)!}(k+2)\sin(k+1)\theta > 0, \qquad\qquad 0 < \theta < \pi.$$

Neither of these inequalities needs an elaborate theory when it comes to a proof, but they would not have been found yet without the general problems to be considered in Lecture 8. And there are other inequalities for sine series which have only been proven by means of the machinery to be developed in these lectures.

Two other results should be mentioned since they provide simple introductions to two of the succeeding lectures.

THEOREM 1.2 (Fejér [6]). *Let* $\sum_{n=1}^{\infty} n|a_n| < \infty$ *and define*

$$f(\theta) = \sum_{n=1}^{\infty} na_n \sin n\theta,$$

$$g(\theta, \varphi) = \sum_{n=1}^{\infty} a_n \sin n\theta \sin n\varphi.$$

Then $f(\theta) \geq 0, 0 \leq \theta \leq \pi$ *if and only if* $g(\theta, \varphi) \geq 0, 0 \leq \theta, \varphi \leq \pi$.
    Proof.

$$f(\theta) = \lim_{\varphi \to 0} \frac{g(\theta, \varphi)}{\varphi}$$

and

(1.32) $$g(\theta, \varphi) = \frac{1}{2} \int_{|\theta - \varphi|}^{\min(\theta + \varphi, 2\pi - \theta - \varphi)} f(x)\, dx.$$

Finally there is a result of Marx [1]:

(1.33) $$\frac{\sin(n+1)\theta}{(n+1)\sin\theta} = \int_{0}^{\pi} \cos n\varphi \, d\mu_\theta(\varphi),$$

where $d\mu_\theta(\varphi) \geq 0, 0 \leq \theta, \varphi \leq \pi$.

A proof similar to the above proof of Theorem 1.1 has been given by Askey and Fitch [2] using

(1.34) $$\frac{d}{d\theta} \frac{\sin(n+1)\theta}{(n+1)(\cos\theta)^{n+1}} = \frac{\cos n\theta}{(\cos\theta)^{n+2}}.$$

LECTURE 2

Jacobi Polynomial Series

The Jacobi polynomials, $P_n^{(\alpha,\beta)}(x)$, provide a class of functions which include all of the basic functions which were used in Lecture 1. These polynomials can be defined by

$$(2.1) \qquad (1 - x)^\alpha (1 + x)^\beta P_n^{(\alpha,\beta)}(x) = \frac{(-1)^n}{2^n n!} \frac{d^n}{dx^n} [(1 - x)^{n+\alpha}(1 + x)^{n+\beta}],$$

or by

$$(2.2) \qquad P_n^{(\alpha,\beta)}(x) = \frac{(\alpha + 1)_n}{n!} {}_2F_1(-n, n + \alpha + \beta + 1; \alpha + 1; (1 - x)/2),$$

where

$$(2.3) \qquad {}_pF_q(\alpha_1, \cdots, \alpha_p; \beta_1, \cdots, \beta_q; x) = \sum_{n=0}^{\infty} \frac{(\alpha_1)_n \cdots (\alpha_p)_n x^n}{(\beta_1)_n \cdots (\beta_q)_n n!}.$$

Initially the assumption that β_i is not a negative integer or zero will be made, but this condition will be removed when the polynomials orthogonal with respect to the binomial and hypergeometric distributions are considered.

The elementary functions in the first lecture arise from

$$(2.4) \qquad \frac{P_n^{(-1/2,-1/2)}(\cos \theta)}{P_n^{(-1/2,-1/2)}(1)} = \cos n\theta,$$

$$(2.5) \qquad \frac{P_n^{(1/2,1/2)}(\cos \theta)}{P_n^{(1/2,1/2)}(1)} = \frac{\sin (n + 1)\theta}{(n + 1) \sin \theta},$$

$$(2.6) \qquad \frac{P_n^{(1/2,-1/2)}(\cos \theta)}{P_n^{(-1/2,1/2)}(1)} = \frac{\sin (n + \frac{1}{2})\theta}{\sin \frac{1}{2}\theta}.$$

Observe that

$$(2.7) \qquad P_n^{(\alpha,\beta)}(1) = \frac{(\alpha + 1)_n}{n!}.$$

Jacobi polynomials are orthogonal on $[-1, 1]$ with respect to $(1 - x)^\alpha (1 + x)^\beta$ when $\alpha, \beta > -1$, and usually this restriction will be assumed. However, many of the formulas hold without this restriction, so occasionally it will be dropped if interesting results are thus obtained. In such cases $P_n^{(\alpha,\beta)}(x)$ will be defined by (2.1), since (2.2) has to be interpreted as a limit when α is a negative integer.

7

The orthogonality conditions are

(2.8) $$\int_{-1}^{1} P_m^{(\alpha,\beta)}(x)P_n^{(\alpha,\beta)}(x)(1-x)^{\alpha}(1+x)^{\beta}\,dx = 0, \qquad\qquad m \neq n,$$

(2.9) $$\int_{-1}^{1} [P_n^{(\alpha,\beta)}(x)]^2(1-x)^{\alpha}(1+x)^{\beta}\,dx = h_n^{\alpha,\beta}$$

$$= \frac{2^{\alpha+\beta+1}}{2n+\alpha+\beta+1}\frac{\Gamma(n+\alpha+1)\Gamma(n+\beta+1)}{\Gamma(n+1)\Gamma(n+\alpha+\beta+1)}.$$

When $f(x) \in L_{\alpha,\beta}^1(-1,1)$, i.e., $f(x)$ is measurable and

(2.10) $$\|f\|_1 = \|f\|_{1,\alpha,\beta} = \int_{-1}^{1} |f(x)|(1-x)^{\alpha}(1+x)^{\beta}\,dx < \infty,$$

then the formal Fourier–Jacobi series of $f(x)$ is

(2.11) $$f(x) \sim \sum_{n=0}^{\infty} \frac{a_n P_n^{(\alpha,\beta)}(x)P_n^{(\alpha,\beta)}(1)}{h_n^{\alpha,\beta}},$$

where

(2.12) $$a_n = \int_{-1}^{1} f(x)\frac{P_n^{(\alpha,\beta)}(x)}{P_n^{(\alpha,\beta)}(1)}(1-x)^{\alpha}(1+x)^{\beta}\,dx.$$

There are many possible ways to normalize Jacobi polynomials and each of the natural ways has advantages and disadvantages. The normalization we have used has two advantages, the symmetry relation

(2.13) $$P_n^{(\alpha,\beta)}(-x) = (-1)^n P_n^{(\beta,\alpha)}(x)$$

is as simple as possible, and it is the classical notation used by Szegö [9] and the authors of the Bateman Project (see Erdélyi et al. [2]). It is convenient to have the Fourier coefficients defined by (2.12) and so the Fourier–Jacobi series has the somewhat unnatural-looking expansion (2.11). It looks a little nicer when

(2.14) $$\frac{P_n^{(\alpha,\beta)}(1)}{h_n^{\alpha,\beta}} = \frac{(2n+\alpha+\beta+1)\Gamma(n+\alpha+\beta+1)}{2^{\alpha+\beta+1}\Gamma(n+\beta+1)\Gamma(\alpha+1)}$$

is used in (2.11), but the apparent simplification is not useful. It is important to realize which constants are important and which ones are not. Coefficients should be written in the way which is easiest to remember. We shall often use

(2.15) $$\frac{P_n^{(\alpha,\beta)}(x)}{P_n^{(\alpha,\beta)}(1)} = {}_2F_1\left(-n, n+\alpha+\beta+1; \alpha+1; \frac{1-x}{2}\right)$$

or

(2.16) $$\frac{P_n^{(\alpha,\beta)}(x)}{P_n^{(\beta,\alpha)}(1)} = (-1)^n {}_2F_1\left(-n, n+\alpha+\beta+1; \beta+1; \frac{1-x}{2}\right),$$

the first because it is one when $x = 1$, the second because it is $(-1)^n$ when $x = -1$ (see (2.4), (2.5), and (2.6)). If the explicit value of $P_n^{(\alpha,\beta)}(1)$ is not being used, then it only confuses the reader (and the author) to use it rather than $P_n^{(\alpha,\beta)}(1)$. As an example, I can never remember the constants in

$$(2.17)\, (n + \alpha + 1)P_n^{(\alpha,\beta)}(x) - (n + 1)P_{n+1}^{(\alpha,\beta)}(x) = \frac{(2n + \alpha + \beta + 2)(1 - x)P_n^{(\alpha+1,\beta)}(x)}{2},$$

while I find it easy to remember

$$(2.18) \qquad \frac{P_n^{(\alpha,\beta)}(x)}{P_n^{(\alpha,\beta)}(1)} - \frac{P_{n+1}^{(\alpha,\beta)}(x)}{P_{n+1}^{(\alpha,\beta)}(1)} = c_n(1 - x)\frac{P_n^{(\alpha+1,\beta)}(x)}{P_n^{(\alpha+1,\beta)}(1)}$$

and the coefficient $c_n = (2n + \alpha + \beta + 2)/(2\alpha + 2)$ can easily be determined by setting $x = -1$. Part of the secret of success in studying and using special functions is to try to remember exactly what is necessary, and nothing more.

All of the formulas which are stated without reference are in Szegö [9], Chapter 4 for Jacobi polynomials and Chapter 5 for Laguerre and Hermite polynomials. They are also usually in Chapter 10 of Erdélyi et al. [2].

Given two functions f, g in $L_{\alpha,\beta}^1$ it is natural to try to set up a formal analogue of convolution so that if a_n and b_n are the Fourier–Jacobi coefficients of f and g respectively (defined by (2.12)), then $h(x) = f*g(x)$ has coefficients given by

$$(2.19) \qquad c_n = a_n b_n = \int_{-1}^{1} h(x)\frac{P_n^{(\alpha,\beta)}(x)}{P_\alpha^{(\alpha,\beta)}(1)}(1 - x)^\alpha(1 + x)^\beta\, dx.$$

To do this form the "generalized translation" of $g(x)$ as the formal series

$$(2.20) \qquad g(x\,;y) \sim \sum_{n=0}^{\infty} b_n\frac{P_n^{(\alpha,\beta)}(x)P_n^{(\alpha,\beta)}(y)}{h_n^{\alpha,\beta}}$$

and define

$$(2.21) \qquad h(x) = \int_{-1}^{1} f(y)g(x\,;y)(1 - y)^\alpha(1 + y)^\beta\, dy.$$

If the b_n are small enough, then the series (2.20) converges absolutely, (2.21) is well-defined, and (2.19) is clearly true; but for general functions f, g only in $L_{\alpha,\beta}^1$ it is not clear that (2.20) and (2.21) make any sense. That they do when $\alpha \geq \beta \geq -\frac{1}{2}$ follows from the following important formulas. When $\alpha = \beta = -\frac{1}{2}$,

$$(2.22) \qquad g(\cos\theta\,;\cos\varphi) = \tfrac{1}{2}[g(\cos(\theta + \varphi)) + g(\cos(\theta - \varphi))]$$

and when $\alpha = \beta > -\frac{1}{2}$,

$$(2.23) \quad g(\cos\theta\,;\cos\varphi) = \frac{\int_0^\pi g(\cos\theta\cos\varphi + \sin\theta\sin\varphi\cos\psi)(\sin\psi)^{2\alpha}\, dx}{\int_0^\pi (\sin\psi)^{2\alpha}\, dx},$$

as can be shown by integrating Gegenbauer's addition formula for ultraspherical polynomials. This will be given in Lecture 4. When $\alpha > \beta > -\frac{1}{2}$ there is a new addition formula for Jacobi polynomials which was found for $\beta = 0$ by Šapiro [1] and independently in the general case by Koornwinder [1], [2], [3], [4], [5], [7], [8]. This formula can be integrated to obtain

$$g(x;y) = \int_0^1 \int_0^\pi g(xy - \tfrac{1}{2}(1-x)(1-y)(1-v^2)$$

(2.24)

$$+ (1-x^2)^{1/2}(1-y^2)^{1/2} v \cos \theta) \, dm_{\alpha,\beta}(\theta, v),$$

where

(2.25) $$dm_{\alpha,\beta}(\theta, v) = \frac{(1-v^2)^{\alpha-\beta-1} v^{2\beta+1}(\sin \theta)^{2\beta} \, d\theta \, dv}{\int_0^1 \int_0^\pi (1-v^2)^{\alpha-\beta-1} v^{2\beta+1}(\sin \theta)^{2\beta} \, d\theta \, dv}, \quad \alpha > \beta > -\tfrac{1}{2}.$$

This formula will be discussed in Lecture 4.

These formulas give a direct definition of $g(x;y)$ when $\alpha \geq \beta \geq -\frac{1}{2}$, a limiting argument in (2.25) is necessary when $\beta = -\frac{1}{2}$, and the norm inequality

(2.26) $$\|f * g\|_1 \leq \|f\|_1 \|g\|_1$$

follows easily. To prove (2.26) and the positivity of the generalized translation operator it is sufficient to show that

$$|xy - \tfrac{1}{2}(1-x)(1-y)(1-v^2) + (1-x^2)^{1/2}(1-y^2)^{1/2} v \cos \theta| \leq 1,$$

$-1 \leq x, y \leq 1, 0 \leq v \leq 1, 0 \leq \theta \leq \pi$. This is a simple exercise in using Cauchy's inequality and the inequality of the arithmetic-geometric mean.

The positivity of the generalized translation operator says that if $\alpha \geq \beta \geq -\frac{1}{2}$ and if

(2.27) $$f(x;1) = \sum_{n=0}^\infty \frac{a_n P_n^{(\alpha,\beta)}(x) P_n^{(\alpha,\beta)}(1)}{h_n} \geq 0, \qquad -1 \leq x \leq 1,$$

then

(2.28) $$f(x;y) = \sum_{n=0}^\infty \frac{a_n P_n^{(\alpha,\beta)}(x) P_n^{(\alpha,\beta)}(y)}{h_n} \geq 0, \qquad -1 \leq x, y \leq 1.$$

When $\alpha = \beta > -\frac{1}{2}$ formula (2.23) is the limiting case of (2.24), just as (2.22) is the limiting case $\alpha = -\frac{1}{2}$ of (2.23). There is a similar limiting case when $\alpha > \beta = -\frac{1}{2}$.

There are a number of senses in which the positivity of (2.27) can be interpreted. Any of the summability methods to be discussed below can be used, or it can be taken in the sense of distributions. All the applications which will be given will concern series which converge absolutely (and usually are only finite series) and in this case the inequalities (2.27) and (2.28) can be taken at face value.

It is not possible at this time to find explicit sums for the Jacobi series analogues of all the series considered in Lecture 1 and obtain their positivity by inspection.

But it is possible for some of the series. The most attractive sum is the analogue of the de la Vallée Poussin sum (1.5). To see how to generalize (1.5) set $x = \cos \theta$. Then $(\cos \frac{1}{2}\theta)^{2n} = (\frac{1}{2}(1 + x))^n$, so a natural analogue of the de la Vallée Poussin sum is

$$(2.29) \qquad \left(\frac{1 + x}{2}\right)^n = \sum_{k=0}^{n} c_{k,n} P_k^{(\alpha,\beta)}(x) P_k^{(\alpha,\beta)}(1).$$

Using orthogonality and integrating by parts k times, (use (2.1)), the coefficients $c_{k,n}$ can be computed:

$$
\left(\frac{1 + x}{2}\right)^n = \sum_{k=0}^{n} c_{k,n} P_k^{(\alpha,\beta)}(x) P_k^{(\alpha,\beta)}(1)
$$

$$(2.30)
\begin{aligned}
= \sum_{k=0}^{n} &\frac{\Gamma(n + \beta + 1)\Gamma(n + 1)(2k + \alpha + \beta + 1)}{\Gamma(k + n + \alpha + \beta + 2)\Gamma(k + \beta + 1)\Gamma(n - k + 1)\Gamma(k + \alpha + 1)} \\
& \times \Gamma(k + \alpha + \beta + 1)\Gamma(\alpha + 1)\Gamma(k + 1) \\
&\cdot P_k^{(\alpha,\beta)}(x) P_k^{(\alpha,\beta)}(1).
\end{aligned}
$$

In this case the generalized translation can be given explicitly:

$$(2.31) \qquad \left(\frac{x + y}{2}\right)^n \frac{P_n^{(\alpha,\beta)}((1 + xy)/(x + y))}{P_n^{(\alpha,\beta)}(1)} = \sum_{k=0}^{n} c_{k,n} P_k^{(\alpha,\beta)}(x) P_k^{(\alpha,\beta)}(y).$$

The positivity for $-1 \leq x, y \leq 1, x + y \neq 0$, is obvious from (2.31), for all the zeros of $P_n^{(\alpha,\beta)}(t)$ lie in $-1 < t < 1$ and $|(1 + xy)/(x + y)| \geq 1$ when $-1 \leq x, y \leq 1$. When $x + y = 0$ a limiting argument shows that the left-hand side is $d_n(1 + xy)^n$ for some d_n and this is strictly positive when $-1 < x < 1$. Formula (2.31) was discovered by Bateman [1] in 1905 and it is by far the nicest formula giving a positive approximate identity for Jacobi series. It was used by Horton [2] to prove the variation diminishing property of the de la Vallée Poussin means. This generalized earlier work of Pólya and Schoenberg [1] for the case $\alpha = \beta = -\frac{1}{2}$.

The next series which was evaluated was the Poisson kernel

$$(2.32)
\begin{aligned}
\sum_{n=0}^{\infty} &\frac{r^n P_n^{(\alpha,\beta)}(x) P_n^{(\alpha,\beta)}(y)}{h_n^{\alpha,\beta}} = \frac{\Gamma(\alpha + \beta + 2)(1 - r)}{2^{\alpha + \beta + 1}\Gamma(\alpha + 1)\Gamma(\beta + 1)(1 + r)^{\alpha + \beta + 2}} \\
&\cdot \sum_{m,n=0}^{\infty} \frac{((\alpha + \beta + 2)/2)_{m+n}((\alpha + \beta + 3)/2)_{m+n}}{(\alpha + 1)_m(\beta + 1)_n m! n!}\left(\frac{a^2}{k^2}\right)^m\left(\frac{b^2}{k^2}\right)^n,
\end{aligned}
$$

where $x = \cos 2\varphi, y = \cos 2\theta, a = \sin \varphi \sin \theta, b = \cos \varphi \cos \theta, k = (r^{1/2} + r^{-1/2})/2$ (see Bailey [3]). The positivity of (2.32) for $-1 \leq x, y \leq 1, \alpha, \beta > -1, 0 \leq r < 1$ is obvious from this formula, since each term on the right-hand side is positive. The classical Poisson kernel (1.3) is also positive for $-1 < r < 0$, but this is now seen to be less fundamental than the positivity for $0 \leq r < 1$. The Poisson kernel in (2.32) is positive for $-1 < r < 1$ only when $\alpha = \beta$. When $\alpha \geq \beta \geq -\frac{1}{2}$ the positivity of the generalized translation operator and the explicit sum

$$\sum_{n=0}^{\infty} \frac{r^n P_n^{(\alpha,\beta)}(x) P_n^{(\alpha,\beta)}(1)}{h_n^{\alpha,\beta}} = \frac{\Gamma(\alpha + \beta + 2)2^{-\alpha-\beta-1}(1-r)}{\Gamma(\alpha + 1)\Gamma(\beta + 1)(1+r)^{\alpha+\beta+2}}$$

(2.33)

$$\cdot {}_2F_1\left(\frac{\alpha + \beta + 2}{2}, \frac{\alpha + \beta + 3}{2}; \beta + 1; \frac{2r(1+x)}{(1+r)^2}\right)$$

have been used to show that (2.32) is nonnegative only for $r_0 \leqq r \leqq 1$, where r_0 is the largest zero (it is negative) of

$${}_2F_1((\alpha + \beta + 2)/2, (\alpha + \beta + 3)/2; \beta + 1; 4r(1 + r)^{-2}) = 0$$

(see Askey [10]). In particular, when $\alpha = \frac{1}{2}$, $\beta = -\frac{1}{2}$, the Poisson kernel (1.13) is nonnegative for $-3 + 2\sqrt{2} \leqq r < 1$ but not for all θ, φ when $r < -3 + 2\sqrt{2}$.

The analogue of the Cesàro means for Jacobi series ((1.2) when $\alpha = \beta = -\frac{1}{2}$, (1.19) when $\alpha = \beta = \frac{1}{2}$, (1.21) when $\alpha = \frac{1}{2}$, $\beta = -\frac{1}{2}$, (1.23) when $\alpha = \beta = 0$, (1.28) when $\alpha = \beta \geqq -\frac{1}{2}$) will be considered in Lecture 8. The results in this case are still incomplete and there are interesting problems here.

The last remaining positive approximate identity in Lecture 1 which has not been discussed is the Jacobi–Weierstrass kernel (1.4). The formal analogue of the sum in (1.4) is

(2.34)
$$\sum_{n=0}^{\infty} \frac{e^{-tn(n+\alpha+\beta+1)} P_n^{(\alpha,\beta)}(x) P_n^{(\alpha,\beta)}(y)}{h_n^{\alpha,\beta}}.$$

The numbers $n(n + \alpha + \beta + 1)$ arise because they are the eigenvalues of the normal differential equation satisfied by $P_n^{(\alpha,\beta)}(x)$:

(2.35) $$\frac{d}{dx}(1 - x)^{\alpha+1}(1 + x)^{\beta+1}\frac{dy(x)}{dx} + n(n + \alpha + \beta + 1)(1 - x)^{\alpha}(1 + x)^{\beta} y(x) = 0,$$

$$y(x) = P_n^{(\alpha,\beta)}(x).$$

No analogue of Jacobi's expression of (1.4) as a product is known for general (α, β) so other methods must be used to prove that (2.34) is positive for $t > 0$. This positivity has been proven for $\alpha, \beta > -1$, $t > 0$, $-1 \leqq x, y \leqq 1$ in Karlin–McGregor [2] as a limiting result of the positivity of a similar sum for Hahn polynomials. Since Hahn polynomials are very interesting and not too much is known about them, they will be defined now:

(2.36) $$Q_n(x; \alpha, \beta, N) = \sum_{k=0}^{n} \frac{(-n)_k(-x)_k(n + \alpha + \beta + 1)_k}{(-N)_k(\alpha + 1)_k k!},$$

$$\alpha, \beta > -1, \quad n = 0, 1, \cdots, N.$$

The sum on the right-hand side is almost a hypergeometric function. The only problem is that one of the denominator parameters is a negative integer, i.e., $-N$. When $x = 0, 1, \cdots, N$ it is possible to consider the right-hand side as

$${}_3F_2(-n, -x, n + \alpha + \beta + 1; -N, \alpha + 1; 1)$$

and to work formally with the hypergeometric function. I know of no instances when trouble arises from using this formula formally when $x, n = 0, 1, \cdots, N$, but if there is ever doubt, then either the series (2.36) or a limiting argument can be used. For example,

$$Q_n(N; \alpha, \beta, N) = \sum_{k=0}^{n} \frac{(-n)_k(n + \alpha + \beta + 1)_k(-N)_k}{(-N)_k(\alpha + 1)_k k!}$$

$$= \sum_{k=0}^{n} \frac{(-n)_k(n + \alpha + \beta + 1)_k}{(\alpha + 1)_k k!} = {}_2F_1(-n, n + \alpha + \beta + 1; \alpha + 1; 1)$$

$$= \frac{(-n - \beta)_n}{(\alpha + 1)_n} = (-1)^n \frac{(\beta + 1)_n}{(\alpha + 1)_n}.$$

Also

$$Q_N(x; \alpha, \beta, N) = \sum_{k=0}^{N} \frac{(-N)_k(N + \alpha + \beta + 1)_k(-x)_k}{(-N)_k(\alpha + 1)_k k!}$$

$$= \sum_{k=0}^{x} \frac{(-x)_k(N + \alpha + \beta + 1)_k}{(\alpha + 1)_k k!}$$

when $x = 0, 1, \cdots, N$; so there is no problem in using the hypergeometric representation when $x = 0, 1, \cdots, N$ with the convention that equal factors in the numerator and denominator are cancelled. However, if x is not one of $0, 1, \cdots, N$, then

$$Q_N(x; \alpha, \beta, N) = \sum_{k=0}^{N} \frac{(-x)_k(N + \alpha + \beta + 1)_k}{(\alpha + 1)_k k!},$$

while the result of cancelling $(-N)_k$ from top and bottom in the ${}_3F_2$ is

$$\sum_{k=0}^{\infty} \frac{(-x)_k(N + \alpha + \beta + 1)_k}{(\alpha + 1)_k k!}$$

and this is not $Q_N(x; \alpha, \beta, N)$. A word of warning: in every case when a denominator parameter is a negative integer, use care in treating hypergeometric functions.

Hahn polynomials are orthogonal on $x = 0, 1, \cdots, N$ with respect to the measure $(\alpha + 1)_x(\beta + 1)_{N-x}/x!(N - x)!$; explicitly

$$\sum_{x=0}^{N} Q_n(x; \alpha, \beta, N)Q_m(x; \alpha, \beta, N) \frac{(\alpha + 1)_x(\beta + 1)_{N-x}N!}{x!(N - x)!(\alpha + \beta + 2)_N}$$

$$= 0, \qquad\qquad m \neq n_y, \quad \alpha, \beta > -1,$$

(2.37)

$$= \frac{1}{\pi_n} = \frac{(-1)^n n!(N + \alpha + \beta + 2)_n(\beta + 1)_n(\alpha + \beta + 1)}{(-N)_n(\alpha + 1)_n(\alpha + \beta + 1)_n(2n + \alpha + \beta + 1)}, \qquad m = n.$$

The normalization of the measure in (2.37) is chosen so that it has mass one. A listing of some of the basic relations for Hahn polynomials is given by Karlin and McGregor [3]. They use a slightly different notation, their polynomials $Q_n(x; \alpha, \beta, N)$

are orthogonal on $x = 0, 1, \cdots, N - 1$. Later Karlin changed his notation and I have used his second notation, as has Gasper [6], [7].

One analogue of (2.34) for Hahn polynomials is

$$(2.38) \qquad \sum_{n=0}^{N} e^{-n(n+\alpha+\beta+1)t} Q_n(x; \alpha, \beta, N) Q_n(y; \alpha, \beta, N) \pi_n$$

and Karlin–McGregor [2] have proven the positivity of (2.38) for $t > 0$, x, y $= 0, 1, \cdots, N$. They use a maximum theorem for finite difference equations. See Gasper [6] for an analogue of the Poisson kernel (2.32) for Hahn polynomials. Gasper proved the positivity by finding a generalization of Bailey's sum (2.32). No analogue of the de la Vallée Poussin kernel has been found (as far as I know no one has tried). There is a fair chance that it will be found.

The generalized translation operator for Hahn polynomials is not a positive operator (see Askey and Gasper [4]). Also Hahn polynomials only behave qualitatively like Jacobi polynomials for n much smaller than N (see Wilson [1], Zaremba [1]). However, they are important, for not only do they contain Jacobi polynomials as limits:

$$(2.39) \qquad \lim_{N \to \infty} Q_n(Nx; \alpha, \beta, N) = \frac{P_n^{(\alpha, \beta)}(1 - 2x)}{P_n^{(\alpha, \beta)}(1)},$$

they contain as limits all of the other classical orthogonal polynomials and Bessel functions. More will be said about these other polynomials in later lectures, so their definitions will be given.

Meixner polynomials

$$(2.40) \qquad M_n(x; \beta, c) = {}_2F_1(-n, -x; \beta; 1 - c^{-1}),$$

$$\beta > 0, \quad 0 < c < 1, \quad z = 0, 1, \cdots.$$

Krawtchouk polynomials

$$(2.41) \qquad K_n(x; p, N) = {}_2F_1(-n, -x; -N; p^{-1}),$$

$$0 < p < 1, \quad x = 0, 1, \cdots, N.$$

Charlier polynomials

$$(2.42) \qquad c_n(x; a) = {}_2F_0(-n, -x; -; -a^{-1}), \qquad a > 0, \quad x = 0, 1, \cdots.$$

Laguerre polynomials

$$(2.43) \qquad L_n^{\alpha}(x) = \frac{(\alpha + 1)_n}{n!} {}_1F_1(-n; \alpha + 1; x), \qquad \alpha > -1.$$

Hermite polynomials

$$(2.44) \qquad H_n(x) = (2x)^n {}_2F_0(-n/2, (-n + 1)/2; -; -x^{-2}).$$

Bessel functions

$$(2.45) \qquad J_\alpha(x) = \frac{(x/2)^\alpha}{\Gamma(\alpha + 1)} {}_0F_1(-; \alpha + 1; -x^2/4).$$

These functions satisfy the following orthogonality relations:

(2.40a) $$\sum_{x=0}^{\infty} M_n(x;\beta,c)M_m(x;\beta,c)c^x(\beta)_x/x! = \frac{n!c^{-n}}{(\beta)_n(1-c)^{\beta}}\delta_{mn},$$

$$0 < c < 1, \quad \beta > 0,$$

(2.41a) $$\sum_{x=0}^{N} K_n(x;p,N)K_m(x;p,N)\binom{N}{x}p^x(1-p)^{N-x} = \frac{n!(1-p)^n}{p^n(-1)^n(-N)_n}\delta_{m,n},$$

$$0 < p < 1,$$

(2.42a) $$\sum_{x=0}^{\infty} c_n(x;a)c_m(x;a)a^x/x! = e^a a^n n!\delta_{m,n}, \qquad a > 0,$$

(2.43a) $$\int_0^{\infty} L_n^{\alpha}(x)L_m^{\alpha}(x)x^{\alpha}\,e^{-x}\,dx = \frac{\Gamma(n+\alpha+1)}{n!}\delta_{m,n}, \qquad \alpha > -1,$$

(2.44a) $$\int_{-\infty}^{\infty} H_n(x)H_m(x)\,e^{-x^2}\,dx = \pi^{1/2}2^n n!\delta_{m,n},$$

(2.45a) $$\int_0^{\infty} J_{\alpha}(xz)J_{\alpha}(yz)z\,dz = \delta(x,y), \qquad \alpha > -1.$$

The last orthogonality relation is the only one which does not exist in the classical sense. It can be looked at in the sense of distributions, or as the limit of a summability process. Since Bessel functions have been more extensively studied than any other set of higher transcendental functions, and since they come from a $_pF_q$ with as few parameters as possible; $p = 0$, $q = 1$; $(_0F_0(-;-;x) = e^x$ and $_1F_0(\alpha;-;x) = (1-x)^{-\alpha})$, it is not surprising that analogues of the Poisson and Weierstrass kernels have been computed. They are

$$\int_0^{\infty} J_{\alpha}(xz)J_{\alpha}(yz)\,e^{-tz}z\,dz = \frac{(2\alpha+1)t(xy)^{\alpha}}{\pi}\int_0^{\pi}\frac{\sin^{2\alpha}\varphi\,d\varphi}{(t^2+x^2+y^2-2xy\cos\varphi)^{\alpha+3/2}},$$

(2.45P) $$\alpha > -\tfrac{1}{2},$$

(2.45W) $$\int_0^{\infty} J_{\alpha}(xz)J_{\alpha}(yz)\,e^{-t^2z^2}z\,dz = \frac{1}{t^2}\exp\left(-(x^2+y^2)/(4t^2)\right)I_{\alpha}\!\left(\frac{xy}{2t^2}\right), \quad \alpha > -1,$$

where

$$I_{\alpha}(t) = \sum_{n=0}^{\infty}\frac{(t/2)^{\alpha+2n}}{\Gamma(n+\alpha+1)n!} = \frac{(t/2)^{\alpha}}{\Gamma(\alpha+1)}\,_0F_1(-;\alpha+1;t^2/4).$$

For the classical polynomials the following sums are known:

$$\sum_{n=0}^{\infty} r^n\frac{(\beta)_n c^n}{n!}M_n(x;\beta,c)M_n(y;\beta,c)$$

(2.40W)

$$= (1-cr)^{-\beta-x-y}(1-r)^{x+y}\,_2F_1\!\left(-x,-y;\beta;\frac{r(1-c)^2}{c(1-r)^2}\right),$$

$$0 \leqq r < 1, \quad x,y = 0,1,\cdots,$$

$$\sum_{n=0}^{z} \frac{(-z)_n}{(-N)_n} \binom{N}{n} p^n (1-p)^{N-n} K_n(x;p,N) K_n(y;p,N)$$

(2.41W)
$$= \frac{(x-N)_z(y-N)_z}{(1-p)^{z-N}(-N)_z(-N)_z} {}_3F_2\left(\begin{matrix} -x,-y,-z \\ 1+N-x-z, 1+N-y-z \end{matrix} ; \frac{p-1}{p} \right),$$

$$x, y, z = 0, 1, \cdots, N,$$

(2.42W) $$\sum_{n=0}^{\infty} r^n \frac{a^n}{n!} c_n(x;a) c_n(y;a) = e^{ar}(1-r)^{x+y} {}_2F_0(-x,-y;-;r/[a(1-r)^2]),$$

$$x, y = 0, 1, \cdots, \quad 0 \le r < 1,$$

(2.43W) $$\sum_{n=0}^{\infty} \frac{r^n n! L_n^\alpha(x) L_n^\alpha(y)}{(\alpha+1)_n}$$

$$= (1-r)^{-1-\alpha} \exp\left[-(x+y)r/(1-r)\right] {}_0F_1\left(-;\alpha+1;\frac{xyr}{(1-r)^2}\right),$$

$$0 \le r < 1,$$

(2.44W) $$\sum_{n=0}^{\infty} \frac{r^n H_n(x) H_n(y)}{2^n n!} = (1-r^2)^{-1/2} \exp\left[(-(x^2+y^2)r^2+2xyr)/(1-r^2)\right],$$

$$-1 < r < 1.$$

In each of these cases it is obvious that the right-hand side is positive, since it has been written as the sum of positive terms.

While the series (2.40W) and (2.42W)–(2.44W) seem to be analogues of the Poisson kernel, in fact they are much more analogous to the Weierstrass kernel. For example, when the limit relation

(2.46) $$\lim_{n \to \infty} n^{-\alpha} L_n^\alpha(x/n) = x^{-\alpha/2} J_\alpha(2x^{1/2})$$

is used on (2.43W) the resulting integral is (2.45W). The reason for this is that all the sums (2.4iW) satisfy parabolic equations, while (2.45P) satisfies an elliptic equation in x and t, in the sense that second order operators in x and first order in t occur in parabolic equations while second order in both x and t occur in elliptic equations. For the discrete polynomials (2.36), (2.40), (2.41), and (2.42) these are difference equations, or differential-difference equations, while for Jacobi, Laguerre, and Hermite polynomials and Bessel functions these are differential equations (which are parabolic and elliptic in the classical sense).

Most of the series given above can be generalized by decoupling the orthogonal polynomials. For example, for Charlier polynomials there is the formula

$$\sum_{x=0}^{\infty} \frac{(abt)^x}{x!} c_n(x;a) c_m(x;b)$$

(2.47)
$$= e^{abt}(1-at)^n(1-bt)^m {}_2F_0(-m,-n;-;t/(1-at)(1-bt)).$$

Gasper has obtained similar formulas for Hahn polynomials and has shown why it is necessary to consider this type of generalization (see Gasper [6], [7]).

There is one other important class of hypergeometric functions which has a generalized orthogonality relation and occurs in many different contexts, from problems with conical symmetry (Mehler [1]) to being spherical harmonics on noncompact two-point homogeneous spaces (see Flensted-Jensen and Koornwinder [1]). These are the Jacobi functions

$$(2.48) \qquad \varphi_\lambda^{(\alpha,\beta)}(t) = {}_2F_1\left(\frac{\alpha + \beta + 1 + i\lambda}{2}, \frac{\alpha + \beta + 1 - i\lambda}{2}; \alpha + 1; -(\sinh t)^2\right).$$

The function

$$u(z) = {}_2F_1(a, b; c; z)$$

satisfies the differential equation

$$(2.49) \qquad z(1 - z)\frac{d^2u}{dz^2} + [c - (a + b + 1)z]\frac{du}{dz} - abu = 0.$$

Contrary to the general impression, Euler [1], not Gauss, first defined ${}_2F_1(a, b; c; x)$ as an infinite series and showed that it satisfied the differential equation (2.49). The coefficients in (2.49) are real when a, b, c are real or when c is real and a and b are complex conjugates. This is a formal reason for the interest in the functions $\varphi_\lambda^{(\alpha,\beta)}(t)$. The applications mentioned above are probably more important reasons for considering these functions.

Various papers of Flensted-Jensen [1], [2], Flensted-Jensen and Koornwinder [1] and Schindler [1], and Vilenkin's book [2] should be consulted for an introduction to these interesting functions. They are more complicated to work with than Jacobi polynomials; for example, there is no analogue to the Christoffel–Darboux formula for the partial sums of the formal reproducing kernel; but this does not mean they should not be studied. It only means that part of the work will be more difficult. Since we shall have more than enough to keep us busy for ten lectures in just considering the classical orthogonal polynomials, these functions will not be considered further. But the reader should not draw a false conclusion from this omission. They are quite important and only a start has been made in their study.

Added in proof. There is another expression for the Poisson type integral (2.45P) for Bessel functions which was given by Hardy [2, p. 46]. He observed that the integral on the right-hand side could be evaluated as an associated Legendre function. The positivity is easy to obtain for $\alpha > -1$ from his formula. A different type of generalization of (1.4) to ultraspherical series and applications of it to probability and statistics is given in Hartman–Watson [1].

Feldheim [2] proved

$$(2.50) \qquad \sum_{n=0}^{\infty} \frac{n!(\alpha + \beta + 1)_n(a)_n(b)_n}{(\alpha + 1)_n(\beta + 1)_n(\alpha + \beta + 1)_{2n}} {}_2F_1(a + n, b + n; \alpha + \beta + 2 + 2n; t)t^n$$

$$\cdot P_n^{(\alpha,\beta)}(x)P_n^{(\alpha,\beta)}(y) = F_4\left(a, b; \alpha + 1, \beta + 1; \frac{t(1 - x)(1 - y)}{4}, \frac{t(1 + x)(1 + y)}{4}\right).$$

It is easy to derive Bailey's sum (2.32) from (2.50). Turán called this interesting paper to my attention.

For Krawtchouk polynomials the following series is interesting:

(2.51)
$$\sum_{n=0}^{N} r^n \binom{N}{n} p^n (1-p)^{N-n} K_n(x;p,N) K_n(y;p,N).$$

See page 42.

LECTURE 3

Fractional Integrals and Hypergeometric Functions

Fejér's proof of (1.22) used (1.1) and Mehler's form of Dirichlet's integral for Legendre polynomials

$$(3.1) \qquad P_n(\cos \theta) = \frac{2}{\pi} \int_\theta^\pi \frac{\sin (n + \frac{1}{2})\varphi \, d\varphi}{[2 \cos \theta - 2 \cos \varphi]^{1/2}}$$

(see Fejér [1], [2]). There are many proofs of (3.1), Hermite–Stieltjes [1, p. 43], Whittaker–Watson [1, 15.231], Dinghas [1], Askey [2], Henrici [2], but the one which is of most interest to us is Bateman's proof [2] using a fractional integration connection between hypergeometric functions.

The following theorem is trivial to prove by integrating the power series one term at a time and using Euler's relation

$$(3.2) \qquad \frac{\Gamma(a)\Gamma(b)}{\Gamma(a + b)} = \int_0^1 t^{a-1}(1 - t)^{b-1} \, dt.$$

THEOREM 3.1. *If $b > a > 0$ and $p \leq q + 1$, then*

$$(3.3) \qquad {}_{p+1}F_{q+1}\left(\begin{matrix} \alpha_1, \cdots, \alpha_p, a \\ \beta_1, \cdots, \beta_q, b \end{matrix}; x\right) = \frac{\Gamma(b)}{\Gamma(b - a)\Gamma(a)}$$

$$\cdot \int_0^1 {}_pF_q\left(\begin{matrix} \alpha_1, \cdots, \alpha_p \\ \beta_1, \cdots, \beta_q \end{matrix}; xt\right) t^{a-1}(1 - t)^{b-a-1} \, dt,$$

where $|x| < 1$ when $p = q + 1$.

When $p = 0$, $q = 1$, and $a = \beta_1$, Theorem 3.1 is Sonine's first integral for Bessel functions,

$$(3.4) \qquad x^{\alpha+\beta+1} J_{\alpha+\beta+1}(x) = \frac{1}{2^\beta \Gamma(\beta + 1)} \int_0^x t^{\alpha+1} J_\alpha(t)(x^2 - t^2)^\beta \, dt, \qquad \beta > -1,$$

and when $p = 2$, $q = 1$, $a = \beta_1$ it is Bateman's integral

$$(3.5) \quad {}_2F_1(a, b; c + \mu; x) = \frac{\Gamma(c + \mu)}{\Gamma(c)\Gamma(\mu)} \int_0^1 y^{c-1}(1 - y)^{\mu-1} {}_2F_1(a, b; c; xy) \, dy,$$

$$c, \mu > 0, \quad -1 < x < 1.$$

Sonine's proof appears in Sonine [1] and Bateman's in Bateman [2]. Surprisingly neither of these proofs uses the simple argument given above. I have been unable

19

to find out who first realized that these results could be proved so easily, but it must have been a folk proof for years before a number of people used it in the 1930's Koshliakov [1] used it in 1926 in the special case $p = q = 1$.

There are a number of interesting and useful integrals connecting Jacobi polynomials which are contained in (3.5). First there is

$$(3.6) \quad (1 - x)^{\alpha + \mu} \frac{P_n^{(\alpha + \mu, \beta - \mu)}(x)}{P_n^{(\alpha + \mu, \beta - \mu)}(1)} = \frac{\Gamma(\alpha + \mu + 1)}{\Gamma(\alpha + 1)\Gamma(\mu)} \int_x^1 (1 - y)^\alpha \frac{P_n^{(\alpha, \beta)}(y)}{P_n^{(\alpha, \beta)}(1)} (y - x)^{\mu - 1} \, dy,$$

$$\alpha > -1, \quad \mu > 0, \quad -1 < x < 1.$$

Then using (2.13) we have

$$(3.7) \quad (1 + x)^{\beta + \mu} \frac{P_n^{(\alpha - \mu, \beta + \mu)}(x)}{P_n^{(\beta + \mu, \alpha - \mu)}(1)} = \frac{\Gamma(\beta + \mu + 1)}{\Gamma(\beta + 1)\Gamma(\mu)} \int_{-1}^x (1 + y)^\beta \frac{P_n^{(\alpha, \beta)}(y)}{P_n^{(\beta, \alpha)}(1)} (x - y)^{\mu - 1} \, dy,$$

$$\beta > -1, \quad \mu > 0, \quad -1 < x < 1.$$

Various transformation formulas of $_2F_1$'s exist, for example,

$$(3.8) \qquad {}_2F_1(a, b; c; x) = (1 - x)^{-a} {}_2F_1(a, c - b; c; x/(x - 1))$$

and they can be used on (3.5). Among the integrals which can be obtained by this method are the following:

$$(3.9)$$
$$\frac{(1 - x)^{\alpha + \mu}}{(1 + x)^{n + \alpha + 1}} \frac{P_n^{(\alpha + \mu, \beta)}(x)}{P_n^{(\alpha + \mu, \beta)}(1)} = \frac{2^\mu \Gamma(\alpha + \mu + 1)}{\Gamma(\alpha + 1)\Gamma(\mu)}$$
$$\cdot \int_x^1 \frac{(1 - y)^\alpha}{(1 + y)^{n + \alpha + \mu + 1}} \frac{P_n^{(\alpha, \beta)}(y)}{P_n^{(\alpha, \beta)}(1)} (y - x)^{\mu - 1} \, dy,$$

$$\alpha > -1, \quad \mu > 0, \quad -1 < x < 1,$$

$$(3.10)$$
$$\frac{(1 + x)^{\beta + \mu}}{(1 - x)^{n + \beta + 1}} \frac{P_n^{(\alpha, \beta + \mu)}(x)}{P_n^{(\beta + \mu, \alpha)}(1)} = \frac{2^\mu \Gamma(\beta + \mu + 1)}{\Gamma(\beta + 1)\Gamma(\mu)}$$
$$\cdot \int_{-1}^x \frac{(1 + y)^\beta}{(1 - y)^{n + \beta + \mu + 1}} \frac{P_n^{(\alpha, \beta)}(y)}{P_n^{(\beta, \alpha)}(1)} (x - y)^{\mu - 1} \, dy,$$

$$\beta > -1, \quad \mu > 0, \quad -1 < x < 1.$$

Erdélyi [1] integrated (3.5) by parts a fractional number of times and obtained

$$(3.11)$$
$${}_2F_1(a, b; c; x) = \frac{\Gamma(c)}{\Gamma(\mu)\Gamma(c - \mu)} \int_0^1 t^{\mu - 1}(1 - t)^{c - \mu - 1}(1 - tx)^{\lambda - a - b}$$
$$\cdot {}_2F_1(\lambda - a, \lambda - b; \mu; tx) {}_2F_1(a + b - \lambda, \lambda - \mu; c - \mu; (1 - t)x/(1 - tx)) \, dt,$$

$$\operatorname{Re} c > \operatorname{Re} \mu > 0, \quad |x| < 1.$$

This formula contains all of the above results as special cases. I was unaware of this and so direct proofs of (3.9) and (3.10) were given in Askey–Fitch [3]. T. P. Higgins called (3.11) to my attention and pointed out some of the interesting special cases it contains. There are others, as Gasper [9] has shown that (3.11) contains the following extension of Mehler's integral:

$$(3.12) \quad \frac{P_n^{(\alpha,\beta)}(\cos\theta)}{P_n^{(\alpha,\beta)}(1)} = \frac{2^{(\alpha+\beta+1)/2}\Gamma(\alpha+1)}{\Gamma(\frac{1}{2})\Gamma(\alpha+\frac{1}{2})}(1-\cos\theta)^{-\alpha}(1+\cos\theta)^{-(\alpha+\beta)/2}$$

$$\cdot \int_0^\theta \frac{\cos\left[k+(\alpha+\beta+1)/2\right]\varphi}{(\cos\varphi-\cos\theta)^{1/2-\alpha}}\,{}_2F_1\left(\frac{\alpha+\beta}{2},\frac{\alpha-\beta}{2};\alpha+\frac{1}{2};\frac{\cos\theta-\cos\varphi}{1+\cos\theta}\right)d\varphi,$$

$$0 < \theta < \pi,\ \mathrm{Re}\,\alpha > -\tfrac{1}{2}.$$

When $\alpha = \beta$ this integral is classical (see Erdélyi et al. [1, p. 177]). A formula equivalent to (3.12) was obtained independently by Koornwinder and used by him to obtain a Paley-Wiener-type theorem for Jacobi polynomials series from the known theorem for cosine series (see Koornwinder [9]). Beurling had used the case $\alpha = \beta = 0$ to prove a similar theorem for Legendre series in a lecture in 1957.

In addition to the linear fractional transformations that ${}_2F_1$'s satisfy, some of them are connected by quadratic transformations. For Jacobi polynomials these are

$$(3.13) \quad \frac{P_{2n}^{(\alpha,\alpha)}(x)}{P_{2n}^{(\alpha,\alpha)}(1)} = \frac{P_n^{(\alpha,-1/2)}(2x^2-1)}{P_n^{(\alpha,-1/2)}(1)},$$

$$(3.14) \quad \frac{P_{2n+1}^{(\alpha,\alpha)}(x)}{P_{2n+1}^{(\alpha,\alpha)}(1)} = \frac{xP_n^{(\alpha,1/2)}(2x^2-1)}{P_n^{(\alpha,1/2)}(1)}.$$

The symmetric Jacobi polynomials, $P_n^{(\alpha,\alpha)}(x)$, arise so often that a different notation is often used,

$$(3.15) \quad C_n^\lambda(x) = \frac{(2\lambda)_n}{(\lambda+\frac{1}{2})_n}P_n^{(\lambda-1/2,\lambda-1/2)}(x), \qquad \lambda \neq 0,-1,\cdots.$$

The reason for the notation is historical; $C_n^\lambda(x)$ has the nice generating function

$$(3.16) \quad (1-2xr+r^2)^{-\lambda} = \sum_{n=0}^\infty C_n^\lambda(x)r^n, \qquad \lambda \neq 0.$$

An extension of this generating function to Jacobi polynomials is

$$(3.17) \quad \sum_{n=0}^\infty \frac{(\alpha+\beta+1)_n P_n^{(\alpha,\beta)}(x)r^n}{(\alpha+1)_n}$$

$$= (1-r)^{-\alpha-\beta-1}\,{}_2F_1\left(\begin{array}{c}\dfrac{\alpha+\beta+1}{2},\dfrac{\alpha+\beta+2}{2}\\[4pt]\alpha+1\end{array};\frac{2r(x-1)}{(1-r)^2}\right)$$

(see Rainville [1, 132 (10)]). The historical reason for the usual normalization of Jacobi polynomials is Jacobi's generating function

$$(3.18) \quad \sum_{n=0}^\infty P_n^{(\alpha,\beta)}(x)r^n = 2^{\alpha+\beta}R^{-1}(1-r+R)^{-\alpha}(1+r+R)^{-\beta},$$

where $R = (1 - 2xr + r^2)^{1/2}$. These two generating functions are analogues for Jacobi polynomials of known Laplace transforms of Bessel functions. The general result for Bessel functions is

(3.19)

$$\int_0^\infty e^{-yt} J_\alpha(xt) t^{\mu-1}\, dt$$

$$= \frac{(x/2y)^\alpha \Gamma(\alpha+\mu)}{y^\mu \Gamma(\alpha+1)}\, {}_2F_1\left(\frac{\alpha+\mu}{2},\, \frac{\alpha+\mu+1}{2}\,;\, \alpha+1\,;\, \frac{-x^2}{y^2}\right)$$

(see Watson [2, 13.2 (2)]). There are four cases in which the ${}_2F_1$ in (3.19) reduces to a more elementary function. They are:

(3.20)
$$\int_0^\infty e^{-yt} J_\alpha(xt) t^{-1}\, dt = \frac{[(x^2+y^2)^{1/2} - y]^\alpha}{\alpha x^\alpha}, \qquad \text{Re } \alpha > 0,$$

(3.21)
$$\int_0^\infty e^{-yt} J_\alpha(xt)\, dt = \frac{[(x^2+y^2)^{1/2} - y]^\alpha}{x^\alpha (x^2+y^2)^{1/2}}, \qquad \text{Re } \alpha > -1,$$

(3.22)
$$\int_0^\infty e^{-yt} J_\alpha(xt) t^\alpha\, dt = \frac{(2x)^\alpha \Gamma(\alpha+\tfrac{1}{2})}{(x^2+y^2)^{\alpha+1/2}\Gamma(\tfrac{1}{2})}, \qquad \text{Re } \alpha > -\tfrac{1}{2},$$

(3.23)
$$\int_0^\infty e^{-yt} J_\alpha(xt) t^{\alpha+1}\, dt = \frac{2y(2x)^\alpha \Gamma(\alpha+\tfrac{3}{2})}{(x^2+y^2)^{\alpha+3/2}\Gamma(\tfrac{1}{2})}, \qquad \text{Re } \alpha > -1.$$

Formula (3.17) contains (3.22) as a limiting case, when

(3.24)
$$\lim_{n\to\infty} n^{-\alpha} P_n^{(\alpha,\beta)}(\cos \theta/n) = (2/\theta)^\alpha J_\alpha(\theta)$$

is used, and (3.18) contains (3.21). Corresponding to (3.23) there is

(3.25)
$$\sum_{n=0}^\infty \frac{(2n+\alpha+\beta+1)\Gamma(n+\alpha+\beta+1)}{\Gamma(n+\beta+1)} P_n^{(\alpha,\beta)}(x) r^n$$

$$= \frac{\Gamma(\alpha+\beta+2)(1-r)}{\Gamma(\beta+1)(1+r)^{\alpha+\beta+2}}\, {}_2F_1\left(\frac{\alpha+\beta+2}{2},\, \frac{\alpha+\beta+3}{2}\,;\, \beta+1\,;\, \frac{2r(1+x)}{(1+r)^2}\right)$$

and to (3.20) there corresponds a recent formula of Carlitz [3]

(3.26)
$$\sum_{n=0}^\infty \frac{\alpha}{n+\alpha} P_n^{(\alpha,-1)}(x) r^n = 2^\alpha (1 - r + R)^{-\alpha}.$$

There does not seem to be a simple generalization of (3.20) to general Jacobi polynomials. This is not too surprising; what is surprising is that so many formulas do generalize. Formula (3.21) is equivalent to an integral which has been useful in queueing theory (see Feller [1] for references). Feller gives a proof of an equivalent integral since he could not find an easy proof in the literature. He starts by proving (3.19) but he then misses a very simple proof by not using Kummer's quadratic transformation

$$_2F_1(a, a + \tfrac{1}{2}; b; x)$$

(3.27)
$$= 2^{2a}(1 + (1 - x)^{1/2})^{-2a}$$

$$\cdot {}_2F_1(2a, 2a - b + 1; b; [1 - (1 - x)^{1/2}]/[1 + (1 - x)^{1/2}])$$

(see Kummer [1] or Erdélyi et al. [1, 2.1.5 (26)]). This immediately gives (3.21), and (3.20) is proven in the same way.

Possibly I should not have included a piece of advertising for hypergeometric functions here (it has nothing to do with fractional integration unless you want to consider Laplace transforms as fractional integrals of infinite order), but I first realized the power of hypergeometric functions when trying to prove (3.9) and (3.10). I was sure they must be true for reasons which will be given below, but was unable to prove them for months, and it was only after learning something about hypergeometric functions that Fitch and I were able to prove them. The reader is probably not aware that we have seen one special case of (3.10) in the first lecture. Recall that we used

(3.28)
$$\frac{\sin (n + 1)\theta}{2(n + 1)(\sin \theta/2)^{2n + 2}} = \int_{\theta/2}^{\pi/2} \frac{\sin (2n + 1)\varphi \, d\varphi}{(\sin \varphi)^{2n + 3}}$$

to prove Turán's Theorem 1.1. This is the special case $\alpha = \tfrac{1}{2}, \beta = -\tfrac{1}{2}, \mu = 1$ of (3.10). In fact (3.10) was found first and then (3.28) was obtained as a special case, and from it

(1.10)
$$\frac{d}{dx} \frac{\sin \alpha x}{\alpha(\sin x)^{\alpha}} = - \frac{\sin (\alpha - 1)x}{(\sin x)^{\alpha + 1}}$$

was suggested. I have no idea how one would suspect that (1.10) holds, or that it was useful, but I was sure that (3.10) held and suspected that it would be useful, since it contained as a special case a result which was equivalent to a useful formula of Feldheim.

When the quadratic transformations (3.13) and (3.14) are used on (3.9) for $\beta = -\tfrac{1}{2}$ and $\beta = \tfrac{1}{2}$ respectively one obtains the known integral of Feldheim [1] and Vilenkin [1],

(3.29)
$$\frac{C_n^{\nu}(\cos \theta)}{C_n^{\nu}(1)} \frac{\sin^{2\nu - 1} \theta}{\cos^{n + 2\lambda + 1} \theta} = \frac{2\Gamma(\nu + \tfrac{1}{2})}{\Gamma(\lambda + \tfrac{1}{2})\Gamma(\nu - \lambda)} \int_0^{\theta} \sin^{2\lambda} \varphi$$

$$\cdot \frac{[\cos^2 \varphi - \cos^2 \theta]^{\nu - \lambda - 1}}{\cos^{n + 2\nu} \varphi} \frac{C_n^{\lambda}(\cos \varphi)}{C_n^{\lambda}(1)} d\varphi,$$

$$0 < \theta < \pi/2, \, \nu > \lambda > -\tfrac{1}{2},$$

with a similar formula for $\pi/2 < \theta < \pi$. An equivalent integral is

$$\frac{C_n^\nu(\cos\theta)}{C_n^\nu(1)} = \frac{2\Gamma(\nu + \frac{1}{2})}{\Gamma(\lambda + \frac{1}{2})\Gamma(\nu - \lambda)} \int_0^{\pi/2} \sin^{2\lambda}\varphi \cos^{2\nu-2\lambda-1}\varphi[1 - \sin^2\theta\cos^2\varphi]^{n/2}$$

(3.30)

$$\cdot \frac{C_n^\lambda(\cos\theta(1 - \sin^2\theta\cos^2\varphi)^{-1/2})\,d\varphi}{C_n^\lambda(1)},$$

$$\nu > \lambda > -\tfrac{1}{2}, \quad 0 \leqq \theta \leqq \pi.$$

Here we have removed the restriction $\lambda \neq 0$ in (3.15), by defining

(3.31)
$$\frac{C_n^0(\cos\theta)}{C_n^0(1)} = \lim_{\lambda\to 0}\frac{C_n^\lambda(\cos\theta)}{C_n^\lambda(1)} = \cos n\theta.$$

Feldheim and Vilenkin each used Sonine's first integral (3.4) and the generating function

(3.32)
$$\sum_{n=0}^{\infty} \frac{C_n^\lambda(\cos\theta)}{C_n^\lambda(1)}\frac{r^n}{n!} = 2^{\lambda-1/2}\Gamma(\lambda + \tfrac{1}{2})\,e^{r\cos\theta}[r\sin\theta]^{1/2-\lambda}J_{\lambda-1/2}(r\sin\theta).$$

While this integral of Feldheim and Vilenkin is useful I was not happy with it. The formula (3.30) was too complicated and (3.29) had a singularity where one should not exist. For the integral (3.29) diverges when $\theta = \pi/2$. I could understand an integral representing Jacobi polynomials diverging when $x = \pm 1$ (or $\theta = 0, \pi$), for they are orthogonal with respect to $(1 - x)^\alpha(1 + x)^\beta$ and $x = \pm 1$ clearly play a distinguished role. However, there should be no difference between $x = 0$ $(\theta = \pi/2)$ and any other point, $-1 < x < 1$. So (3.29) was probably not the basic formula and one should try to find a more fundamental one which does not have the strange singularity when $\theta = \pi/2$. After using the quadratic transformations (3.13) and (3.14) on (3.29) I obtained (3.9) for $\beta = \pm\frac{1}{2}$, and surprisingly the formulas were identical except for the Jacobi polynomials. This strongly suggested that (3.9) held for all $\beta > -1$.

Feldheim [1] gave one interesting application of (3.30). As was remarked above Fejér proved

(1.22)
$$\sum_{k=0}^{n} P_k(x) > 0, \qquad\qquad -1 < x \leqq 1,$$

by using (3.1) on (1.1). Since $P_n(x) = C_n^{1/2}(x)$, (3.30) can be summed to give

(3.33)
$$\sum_{k=0}^{n} \frac{C_k^\nu(\cos\theta)}{C_k^\nu(1)} = \frac{2\Gamma(\nu + \frac{1}{2})}{\Gamma(\nu - \frac{1}{2})}\int_0^{\pi/2} \sin\varphi\cos^{2\nu-2}\varphi$$

$$\cdot \sum_{k=0}^{n}[1 - \sin^2\theta\cos^2\varphi]^{k/2}P_k(\cos\theta(1 - \sin^2\theta\cos^2\varphi)^{-1/2})\,d\varphi,$$

$$\nu > \tfrac{1}{2}.$$

A summation by parts and (1.22) show that

$$(3.34) \qquad \sum_{k=0}^{n} a_k P_k(x) > 0, \qquad\qquad -1 < x \leq 1,$$

when $a_k \geq a_{k+1} \geq 0$, $a_0 > 0$, $k = 0, 1, \cdots, n - 1$. If $a_k = [1 - \sin^2 \theta \cos^2 \varphi]^{k/2}$, then $a_k \geq a_{k+1} \geq 0$ so the sum under the ·integral in (3.33) is positive. Thus Feldheim concluded that

$$(3.35) \qquad \sum_{k=0}^{n} \frac{C_k^v(\cos \theta)}{C_k^v(1)} > 0, \qquad\qquad 0 \leq \theta < \pi, \quad v \geq \tfrac{1}{2}.$$

When $v = 1$ this is the Fejér-Jackson-Gronwall inequality (1.6).

Feldheim's trick of proving (3.34) by summing by parts will not always work on the series we wish to consider so a more complicated argument must be used. The Poisson kernel which was discussed in Lecture 2 will be the substitute. It will be used to prove the following theorem.

THEOREM 3.2. *If $\gamma > \alpha > -1$ and*

$$f(x) = \sum_{k=0}^{n} a_k \frac{P_k^{(\alpha,\alpha)}(x)}{P_k^{(\alpha,\alpha)}(1)} \geq 0, \qquad\qquad -1 \leq x \leq 1,$$

then

$$g(y) = \sum_{k=0}^{n} a_k \frac{P_k^{(\gamma,\gamma)}(y)}{P_k^{(\gamma,\gamma)}(1)} > 0, \qquad\qquad -1 < y < 1,$$

unless $a_k \equiv 0$, $k = 0, 1, \cdots, n$.

Proof. Use of the Feldheim-Vilenkin formula gives

$$g(\cos \theta) = \frac{2\Gamma(\gamma + 1)}{\Gamma(\gamma - \alpha)\Gamma(\alpha + 1)} \int_0^{\pi/2} \sin^{2\alpha+1} \varphi \cos^{2\gamma-2\alpha-1} \varphi$$

$$\cdot \sum_{k=0}^{n} a_k [1 - \sin^2 \theta \cos^2 \varphi]^{k/2} \frac{P_k^{(\alpha,\alpha)}(\cos \theta(1 - \sin^2 \theta \cos^2 \varphi)^{-1/2})}{P_k^{(\alpha,\alpha)}(1)} \, d\varphi.$$

Observe that $[1 - \sin^2 \theta \cos^2 \varphi] < 1$ for $0 < \theta < \pi$, $0 \leq \varphi < \pi/2$, so the strict positivity of the Poisson kernel (which follows from Bailey's formula (2.32)) implies that the series under the integral is strictly positive for $0 < \theta < \pi$, $0 \leq \varphi < \pi/2$ unless $f(x) \equiv 0$, $-1 \leq x \leq 1$. This only happens when $a_k = 0$, $k = 0, 1, \cdots, n$.

A slightly more complicated argument can be given to prove the following.

THEOREM 3.3. *If $\gamma > \alpha > -1$, then*

$$(3.36) \qquad \frac{P_n^{(\gamma,\gamma)}(x)}{P_n^{(\gamma,\gamma)}(1)} = \int_{-1}^{1} \frac{P_n^{(\alpha,\alpha)}(y)}{P_n^{(\alpha,\alpha)}(1)} \, d\mu_x(y), \qquad\qquad -1 \leq x \leq 1,$$

where $d\mu_x(y) \geq 0$. When $-1 < x < 1$, $d\mu_x(y)$ is absolutely continuous, it is strictly positive and finite for $-1 < x, y < 1$, $x \neq y$, and is given by

$$(3.37) \qquad k_{\alpha,\gamma}(x, y) = \sum_{n=0}^{\infty} \frac{P_n^{(\alpha,\alpha)}(1)}{P_n^{(\gamma,\gamma)}(1)} [h_n^{\alpha,\alpha}]^{-1} P_n^{(\gamma,\gamma)}(x) P_n^{(\alpha,\alpha)}(y).$$

This series converges for $-1 < x < 1, -1 < y < 1, x \neq y$. *When* $x^2 = 1, d\mu_x(y)$
is a unit mass at $y = x$.

Most of Theorem 3.3 was proved in Askey [3] and Bingham [1]. The strict
positivity follows from the strict positivity of the Poisson kerenel and the convergence follows from asymptotic formulas (see Szegö [9, Chap. 8]). The special
case when $\gamma = \frac{1}{2}, \alpha = -\frac{1}{2}$ was stated by Marx [1] as a problem and solved by
Koschmieder and Stroman [1]. Seidel and Szász [1] extended this to $\gamma > \alpha = -\frac{1}{2}$.

Further integral connections similar to (3.36) are contained in the following
theorem.

THEOREM 3.4. *Jacobi polynomials are connected by the integral relation*

$$(3.38) \qquad \frac{P_n^{(\gamma,\delta)}(x)}{P_n^{(\delta,\gamma)}(1)} = \int_{-1}^{1} \frac{P_n^{(\alpha,\beta)}(y)}{P_n^{(\beta,\alpha)}(1)} \, d\mu_x(y),$$

where $d\mu_x(y) = d\mu_x(y; \alpha, \beta, \gamma, \delta)$ *is a nonnegative measure when*

(i) $\gamma + \delta = \alpha + \beta, \gamma < \alpha, \beta > -1$;

(ii) $\gamma = \alpha > -1, \delta > \beta > -1$;

(iii) $\gamma \leq \alpha, \gamma + \delta \geq \alpha + \beta, \alpha, \beta > -1$;

(iv) $\gamma = \delta, \alpha = \beta, \gamma > \alpha > -1$;

(v) $\gamma = \delta + 1, \alpha = \beta + 1, \delta > \beta > -1$.

In all these cases the measure is absolutely continuous for $-1 < x < 1, -1 < y < 1$,
$x \neq y$, *and in cases* (ii), (iv), *and* (v) *it is strictly positive for* $-1 < x < 1, -1 < y < 1$,
$x \neq y$.

Remark. Part (i) is Bateman's integral (3.7). Part (ii) is proved from (3.10) using
the positivity of the Poisson kernel as in Theorem 3.2. Part (iii) is just a combination
of (i) and (ii), while (iv) is Theorem 3.3. Part (v) follows from (iv) by use of a contiguous relation (see Askey [3]).

The following conjecture is probably true, probably hard to prove, and it
would be useful.

CONJECTURE 3.1. *The conclusion of Theorem 3.4 holds when*

$$(\gamma, \delta) = (\alpha + \mu, \beta + \mu), \quad \mu > 0, \quad \alpha \geq \beta > -1.$$

The reason for the strange normalization in (3.38) will become clear in the next
theorem and the application given after this theorem.

THEOREM 3.5. *If* $\alpha, \beta > -1$,

$$(3.39) \qquad f(x) = \sum_{k=0}^{\infty} a_k \frac{P_k^{(\alpha,\beta)}(x)}{P_k^{(\beta,\alpha)}(1)} \geq 0, \qquad -1 \leq x \leq 1,$$

and

$$(3.40) \qquad \sum_{k=0}^{\infty} |a_k|[(k+1)^{1/2-\beta} + 1 + (k+1)^{\alpha-\beta}] < \infty,$$

then

$$(3.41) \qquad g(y) = \sum_{k=0}^{\infty} a_k \frac{P_k^{(\gamma,\delta)}(y)}{P_k^{(\delta,\gamma)}(1)} \geq 0, \qquad -1 \leq y \leq 1,$$

when (γ, δ) *and* (α, β) *satisfy any of the conditions in Theorem* 3.3. *In cases* (ii), (iv) *and* (v), $g(y) > 0$, $-1 < y < 1$ *unless* $f(x) \equiv 0$, $-1 \leqq x \leqq 1$.

Remark. The condition (3.40) is the easiest condition which assures the absolute convergence (and therefore uniform convergence) of (3.39) so that (3.38) can be applied. It can be weakened, but we will not need a stronger result since Theorem 3.5 will only be applied to polynomials, and in this case (3.40) is trivially satisfied. When $\alpha = \gamma = \delta = \frac{1}{2}$, $\beta = -\frac{1}{2}$, this theorem reduces to Theorem 1.1 of Turán.

Theorem 3.5 can be applied to the series (3.35) which was considered above, but it is more efficient to apply it directly to Fejér's sum

$$(3.42) \qquad \sum_{k=0}^{n} \frac{P_k^{(1/2,-1/2)}(\cos \theta)}{P_k^{(-1/2,1/2)}(1)} = \sum_{k=0}^{n} \frac{\sin (k + \frac{1}{2})\theta}{\sin \theta/2} = \frac{1 - \cos (n + 1)\theta}{2 \sin^2 \theta/2} \geqq 0.$$

Applying (v) and then (i) leads to

$$(3.43) \qquad \sum_{k=0}^{n} \frac{P_k^{(\alpha,\beta)}(x)}{P_k^{(\beta,\alpha)}(1)} > 0,$$

$$-1 < x \leqq 1, \quad n = 0, 1, \cdots, \quad -\beta \leqq \alpha \leqq \beta + 1, \quad \beta > -\frac{1}{2}.$$

The reason for the strange normalization is now clear, since the series in (3.43) vanishes for $x = -1$ when n is odd. Later on we shall return to (3.43) and mention an application to proving the positivity of some Cotes' numbers, as well as extend it to other values of (α, β) and generalize it by adding other coefficients. The reader who examines the results of Askey [3] will see that I was unaware of most of these applications even though I had most of the integrals referred to in Theorem 3.4. And somewhat earlier in Askey–Fitch [3] I was completely unaware of the "right" generalization of Feldheim's sum (3.35). The sum considered there was

$$(3.44) \qquad \sum_{k=0}^{n} \frac{P_k^{(\alpha,\beta)}(x)}{P_k^{(\alpha,\beta)}(1)}.$$

This sum can only be nonnegative for $\alpha \geq \beta$ (let $x = -1$ for n odd) and when $\alpha > |\beta|$ it is strictly positive for $-1 \leq x \leq 1$. Sums which vanish at a point and are positive elsewhere are usually deeper than those which are positive everywhere. In this case a summation by parts shows that if (3.43) holds for (α, β) and $\alpha \geq \beta$, then (3.44) is positive for $-1 < x \leq 1$. So (3.43) is more fundamental than (3.44). The reason that (3.43) was missed in Askey–Fitch [3] is that we did not apply it, and so did not see that it was too weak to be very useful. As was pointed out in Askey [7], I discovered (3.43) by trying to use (3.44) in a way similar to the way Szegö [1] used the special case $\alpha = \beta = 0$ to prove a theorem on Cesàro summability. This is typical of what happens when you try to generalize a formula or theorem by adding a new parameter by formally generalizing a previous result. It can usually be done in several ways and an application will often direct your attention to the more fruitful one, while formal generalizations without applications will often go in easier but less useful directions.

LECTURE 4

Addition Formulas

As was suggested in the first lecture, everything we will be doing in these lectures is just an appropriate generalization of known properties of sines and cosines or exponentials to more general functions. The most striking property of e^x is

(4.1) $$e^{x+y} = e^x \cdot e^y,$$

and formulas of a similar type are called addition formulas. For sines and cosines there are the formulas

(4.2) $$\sin(x + y) = \sin x \cos y + \cos x \sin y,$$

(4.3) $$\cos(x + y) = \cos x \cos y - \sin x \sin y.$$

The addition formula of e^x has been extended to locally compact abelian groups, and a very interesting theory of harmonic analysis has been developed in this context (see Rudin [1]).

There are a number of generalizations of these formulas. For example, elliptic functions satisfy an addition formula of the type

(4.4) $$\mathscr{P}(u + v) = A(\mathscr{P}(u), \mathscr{P}(v)),$$

where $A(r, s)$ is an algebraic function. Weierstrass proved that the only solutions of (4.4) are algebraic functions, algebraic functions of e^{ciu} for some constant c, or algebraic functions of elliptic functions (see Copson [1, p. 363] for references).

There are two ways to look at this theorem, either to give up looking for addition formulas, or to realize that further addition formulas are going to be more complicated. Fortunately there was an addition formula which was quite old before Weierstrass gave his theorem. This is Laplace's expression for $P_n(\cos\theta\cos\varphi + \sin\theta\sin\varphi\cos\psi)$ (see Laplace [1]). Notice the date of 1782 about one hundred years before the first published proof of Weierstrass' theorem. Explicitly it is

$$P_n(\cos\theta\cos\varphi + \sin\theta\sin\varphi\cos\psi) = P_n(\cos\theta)P_n(\cos\varphi)$$

(4.5)
$$+ 2\sum_{m=1}^{n} \frac{(n-m)!}{(n+m)!} P_n^m(\cos\theta)P_n^m(\cos\varphi)\cos m\psi,$$

where $P_n^m(x)$ is the associated Legendre function which is defined by

(4.6) $$P_n^m(x) = (-1)^m(1 - x^2)^{m/2}\frac{d^m}{dx^m}P_n(x), \qquad -1 < x < 1.$$

29

As long as one is only working with Legendre polynomials, then the notation of the associated Legendre functions is quite useful and compact. However, when the next step was taken by Gegenbauer in 1875 [2] it became clear that this was not really the right way to write addition formulas. Gegenbauer proved that

$$C_n^\lambda(\cos\theta\cos\varphi + \sin\theta\sin\varphi\cos\psi) = C_n^\lambda(\cos\theta)C_n^\lambda(\cos\varphi)$$

(4.7)

$$+ \sum_{k=1}^n a_{k,n}^\lambda(\sin\theta)^k C_{n-k}^{\lambda+k}(\cos\theta)(\sin\varphi)^k C_{n-k}^{\lambda+k}(\cos\varphi)C_k^{\lambda-1/2}(\cos\psi),$$

with

(4.8)
$$a_{k,n}^\lambda = \frac{\Gamma(2\lambda-1)2^{2k}[\Gamma(k+\lambda)]^2(n-k)!(2k+2\lambda-1)}{[\Gamma(\lambda)]^2\Gamma(n+k+2\lambda)}.$$

This formula reduces to (4.5) when the limit as $\lambda \to \frac{1}{2}$ is taken, since

(4.9)
$$\lim_{\lambda\to 0}\frac{(n+\lambda)}{\lambda}C_n^\lambda(\cos\theta) = \begin{cases} 1 & \text{for } n = 0, \\ 2\cos n\theta & \text{for } n = 1, 2, \cdots. \end{cases}$$

Also observe that (4.7) reduces to (4.3) when $\psi = 0$ and the limit $\lambda \to 0$ is taken.

There are many proofs of Gegenbauer's addition formula, Gegenbauer [2], Nielson [1], Hobson [1], Herglotz (in Erdélyi et al. [2; Chap. 11]), Henrici [1], Robin [1, Chap. 7], Müller [1], Vilenkin [2], Manocha [1], Carlson [1], Koornwinder [4]. Some of the proofs are analytic and hold for all λ; others are group theoretic in nature and hold for $\lambda = j/2, j = 0, 1, \cdots$, and the general result is obtained by applying Carlson's uniqueness theorem for functions analytic in a half-plane. See Boas [1, Chap. 9] for this theorem. It is also possible to obtain (4.7) from (4.5) by differentiating with respect to ψ to obtain (4.7) for $\lambda = 1, 2, \cdots$, and then using Carlson's theorem. It would be hard to overemphasize the importance of this addition formula. Among the formulas which are contained in it is Gegenbauer's product formula

$$\frac{C_n^\lambda(\cos\theta)C_n^\lambda(\cos\varphi)}{C_n^\lambda(1)} = c_\lambda\int_0^\pi C_n^\lambda(\cos\theta\cos\varphi + \sin\theta\sin\varphi\cos\psi)(\sin\psi)^{2\lambda-1}\,d\psi,$$

(4.10) $\lambda > 0,$

where

(4.11)
$$c_\lambda^{-1} = \int_0^\pi (\sin\psi)^{2\lambda-1}\,d\psi,$$

and Gegenbauer's generalization of the Laplace integral representation of Legendre polynomials:

(4.12)
$$\frac{C_n^\lambda(x)}{C_n^\lambda(1)} = c_\lambda\int_0^\pi [x + (x^2-1)^{1/2}\cos\psi]^n\sin^{2\lambda-1}\psi\,d\psi,\qquad \lambda > 0.$$

Observe that (4.10) and (4.12) reduce to well-known formulas for cosines when the limit $\lambda \to 0$ is taken. They become

(4.13) $$\cos n\theta \cos n\varphi = \tfrac{1}{2}[\cos n(\theta + \varphi) + \cos n(\theta - \varphi)]$$

and

(4.14) $$\cos n\theta = \tfrac{1}{2}[e^{in\theta} + e^{-in\theta}],$$

respectively.

Formula (4.10) gives the positivity of the generalized translation operator for ultraspherical series which was mentioned in Lecture 2. The Laplace-type integral (4.12) has been used to study axially symmetric potentials (see Gilbert [1] for details and further references). So it was particularly unfortunate that the corresponding addition formula for Jacobi polynomials was unknown. This formula has recently been found by Šapiro [1] for $P_n^{(\alpha,0)}(x)$ and independently by Koornwinder [1] in the general case $P_n^{(\alpha,\beta)}(x)$. As often happens, once one proof has been found then others follow shortly, and there are now three distinct proofs, Koornwinder [2], [3], [4], [5], [7] and Koornwinder has announced another one using orthogonal polynomials in three variables. This last is probably the easiest proof which can be found. Each of the proofs is interesting and should be studied in detail by anyone who wants to do research in this area. The algebraic proofs are probably the most important, since addition formulas looked at from this point of view are natural generalizations of the rotation property of the circle,

(4.15) $$e^{i\theta} \cdot e^{i\varphi} = e^{i(\theta + \varphi)}.$$

Also the first proof was found by algebraic methods. Unfortunately I do not have the knowledge which would be required to give a good summary of one of the algebraic proofs in the time we have here, and so I shall give part of an analytic proof of the following formula:

$$P_n^{(\alpha,\beta)}(\tfrac{1}{2}(1 + x)(1 + y) + \tfrac{1}{2}(1 - x)(1 - y)r^2 + (1 - x^2)^{1/2}(1 - y^2)^{1/2}r \cos \varphi - 1)$$

$$(k + m + \alpha)(k - m + \beta)(n + \alpha + \beta + 1)_k$$

$$= \sum_{k=0}^{n} \sum_{m=0}^{k} \frac{\cdot (2\beta + 1)_{k-m}(n - m + \beta + 1)_m(n - k)!}{2^{2k-1}(k + \alpha)(k - m + 2\beta)(\beta + 1)_k(\beta + \tfrac{1}{2})_{k-m}(k + \alpha + 1)_{n-k+m}}$$

(4.16) $$\cdot (1 - x)^{(k+m)/2}(1 + x)^{(k-m)/2} P_{n-k}^{(\alpha+k+m,\beta+k-m)}(x)$$

$$\cdot (1 - y)^{(k+m)/2}(1 + y)^{(k-m)/2} P_{n-k}^{(\alpha+k+m,\beta+k-m)}(y)$$

$$\cdot P_m^{(\alpha-\beta-1,\beta+k-m)}(2r^2 - 1)r^{k-m} P_{k-m}^{(\beta-1/2,\beta-1/2)}(\cos \varphi).$$

Since this formula is so complicated a few consequences should be given to show that it is useful. Integrate with respect to the measure $(1 - r^2)^{\alpha - \beta - 1} r^{2\beta + 1}$ $\cdot (\sin \varphi)^{2\beta} \, dr \, d\varphi$ and use the orthogonality of Jacobi polynomials to obtain

$$\frac{P_n^{(\alpha,\beta)}(x) P_n^{(\alpha,\beta)}(y)}{P_n^{(\alpha,\beta)}(1)}$$

(4.17)
$$= \int_0^\pi \int_0^1 P_n^{(\alpha,\beta)}(\tfrac{1}{2}(1 + x)(1 + y) + \tfrac{1}{2}(1 - x)(1 - y)r^2$$

$$+ (1 + x^2)^{1/2}(1 - y^2)^{1/2} r \cos \varphi - 1) \, dm(r, \varphi),$$

where

(4.18) $dm(r, \varphi) = dm_{\alpha,\beta}(r, \varphi) = c_{\alpha,\beta}(1 - r^2)^{\alpha - \beta - 1} r^{2\beta + 1}(\sin \varphi)^{2\beta} \, dr \, d\varphi$

and

(4.19) $c_{\alpha,\beta}^{-1} = \int_0^\pi \int_0^1 (1 - r^2)^{\alpha - \beta - 1} r^{2\beta + 1}(\sin \varphi)^{2\beta} \, dr \, d\varphi, \qquad \alpha > \beta > -\tfrac{1}{2}.$

This product formula gives the explicit formula for the generalized translation operator for Jacobi series when $\alpha > \beta > -\tfrac{1}{2}$, and for $\alpha > \beta = -\tfrac{1}{2}$ by a limiting argument. Gasper [4] has obtained an equivalent formula to (4.17), but it was written in a more complicated way. Divide by y^n and let $y \to \infty$ in (4.17) to obtain the Laplace-type integral

(4.20) $\dfrac{P_n^{(\alpha,\beta)}(x)}{P_n^{(\alpha,\beta)}(1)} = \displaystyle\int_0^\pi \int_0^1 \left[\dfrac{1 + x - (1 - x)r^2}{2} + i\sqrt{1 - x^2} \, r \cos \theta \right]^n dm_{\alpha,\beta}(r, \theta),$

$\alpha > \beta > -\tfrac{1}{2}$. Each of these formulas reduces to the corresponding formula of Gegenbauer when $\beta \to a$ and to the formula which follows from Gegenbauer's formula by means of the quadratic transformation (3.13) when $\beta \to -\tfrac{1}{2}$.

Our proof of (4.16) will start with a proof of (4.20), then go to (4.17) and finally an indication will be given of how to prove (4.16) from (4.17). To prove (4.20) we start even further back and recall Gegenbauer's formula (4.12), which we now write as

(4.21) $\dfrac{P_n^{(\alpha,\alpha)}(x)}{P_n^{(\alpha,\alpha)}(1)} = c_{\alpha + 1/2} \displaystyle\int_0^\pi [x + (x^2 - 1)^{1/2} \cos \psi]^n \sin^{2\alpha} \psi \, d\psi, \qquad \alpha > -\tfrac{1}{2}.$

Simple proofs are given in Szegö [7, (4.10.3)] and Gegenbauer [3]. Gegenbauer's proof is also given in Askey [13] since it appeared in a journal which not all libraries have and it is so simple. Use (4.21) in (3.9) to obtain

$$\frac{P_n^{(\alpha,\beta)}(x)}{P_n^{(\alpha,\beta)}(1)} = 2^{\alpha - \beta - 1} c_{\alpha,\beta} \int_x^1 \int_0^\pi \frac{(1 - y)^\beta (1 + x)^{n + \beta + 1}}{(1 - x)^\alpha (1 + y)^{n + \alpha + 1}} (y - x)^{\alpha - \beta - 1}$$

(4.22)

$$\cdot [y + i(1 - y^2)^{1/2} \cos \theta]^n (\sin \theta)^{2\beta} \, d\theta \, dy.$$

Then change variables, using

$$r^2 = \frac{(1 - y)(1 + x)}{(1 + y)(1 - x)}$$

to obtain (4.20).

Bateman [1] proved the following formula

(4.23) $$\frac{P_n^{(\alpha,\beta)}(s)}{P_n^{(\alpha,\beta)}(1)} \frac{P_n^{(\alpha,\beta)}(t)}{P_n^{(\alpha,\beta)}(1)} = \sum_{k=0}^{n} b_{k,n}(s+t)^k \frac{P_k^{(\alpha,\beta)}((1+st)/(s+t))}{P_k^{(\alpha,\beta)}(1)},$$

where $b_{k,n}$ is defined by (4.23) when $t = 1$,

(4.24) $$\frac{P_n^{(\alpha,\beta)}(s)}{P_n^{(\alpha,\beta)}(1)} = \sum_{k=0}^{n} b_{k,n}(s+1)^k.$$

Bateman's second proof [3] uses the fact that both sides of (4.23) satisfy the same partial differential equation. For the left-hand side it is

$$D_s^{\alpha,\beta} f(s,t) = D_t^{\alpha,\beta} f(s,t),$$

where

$$D_s^{\alpha,\beta} = (1-s)^{-\alpha}(1+s)^{-\beta}\frac{\partial}{\partial s}(1-s)^{\alpha+1}(1+s)^{\beta+1}\frac{\partial}{\partial s}$$

(see Koornwinder [5] for an extended discussion of the types of changes of variable in the biaxially symmetric potential equation which lead to similar formulas, not only for Jacobi polynomials, but also for Bessel functions).

If (4.20) is used in the right-hand side of (4.23), then a simplification leads immediately to (4.17). To complete the proof to (4.16) Koornwinder [7] integrated (4.17) by parts a number of times. He had to derive some new differentiation formulas for hypergeometric functions to be able to carry out the right integrations by parts. The following formulas are typical of the ones he has derived:

(4.25) $$\left(\frac{d^2}{dx^2} + \frac{2c-1}{x}\frac{d}{dx}\right){}_2F_1(a,b;c;x^2) = 4ab{}_2F_1(a+1,b+1;c;x^2),$$

(4.26)
$$\left(\frac{d^2}{dx^2} + \frac{2\beta+1}{x}\frac{d}{dx}\right)[(1-x^2)^{\alpha+2}P_{n-1}^{(\alpha+2,\beta)}(2x^2-1)]$$
$$= 4n(n+\alpha+1)(1-x^2)^{\alpha}P_n^{(\alpha,\beta)}(2x^2-1),$$

(4.27)
$$2^{2n}n!(n+\alpha+1)_n(1-x^2)^{\alpha}P_n^{(\alpha,\beta)}(2x^2-1) = \left(\frac{d^2}{dx^2} + \frac{2\beta+1}{x}\frac{d}{dx}\right)(1-x^2)^{2n+\alpha}$$

(see Koornwinder [7] for the details of the rest of the reduction, since it is quite computational).

Recall Fejér's result which was given in Lecture 1:

(4.28) $$\sum_{n=1}^{\infty} a_n \sin n\theta \sin n\varphi \geqq 0, \qquad\qquad 0 \leqq \varphi, \theta \leqq \pi,$$

if and only if

(4.29) $$\sum_{n=1}^{\infty} na_n \sin n\theta \geqq 0, \qquad\qquad 0 \leqq \theta \leqq \pi.$$

This can be extended to Jacobi polynomials in the following way.

THEOREM 4.1. *Let* $\alpha \geq \beta$ *and either* $\beta \geq -\frac{1}{2}$ *or* $\alpha \geq -\beta, \beta > -1$, *and assume* $\sum_{n=0}^{\infty} |a_n| < \infty$. *Then*

$$
(4.30) \qquad f(x, y) = \sum_{n=0}^{\infty} a_n \frac{P_n^{(\alpha,\beta)}(x)}{P_n^{(\alpha,\beta)}(1)} \frac{P_n^{(\alpha,\beta)}(y)}{P_n^{(\alpha,\beta)}(1)} \geq 0, \qquad -1 \leq x, y \leq 1,
$$

if and only if

$$
(4.31) \qquad f(x, 1) \geq 0, \qquad -1 \leq x \leq 1.
$$

When $\alpha \geq \beta \geq -\frac{1}{2}$ this is an immediate consequence of (4.17). Gasper [5] has proved the remaining case when $-1 < \beta < -\frac{1}{2}$. If the assumption $\sum_{n=0}^{\infty} |a_n| < \infty$ is dropped, then the theorem is still correct when nonnegativity is interpreted in the sense of distributions. Fejér's result mentioned above is just the case $\alpha = \beta = \frac{1}{2}$. The case $\alpha = \frac{1}{2}, \beta = -\frac{1}{2}$ can be proved in an elementary way, and the case $\alpha = \beta = -\frac{1}{2}$ is trivial. The other cases seem to need a more complicated argument. When $\alpha = \beta$ Weinberger [1] proved Theorem 4.1 from a maximum theorem for hyperbolic equations. In fact he found this maximum theorem so as to be able to prove Theorem 4.1 without using Gegenbauer's formula (4.10). Koornwinder has proved Theorem 4.1 when $\alpha \geq \beta \geq -\frac{1}{2}$ in a similar way, but his proof is not yet published. Gasper [5] has shown the conditions on (α, β) in Theorem 4.1 are best possible. For the Jacobi functions $\varphi_\lambda^{(\alpha,\beta)}(t)$, Chébli [1] has proven the analogue of Theorem 4.1. It holds for $\alpha \geq \beta > -1, \alpha + \beta \geq -1$.

In the ultraspherical case a slight refinement has been given.

THEOREM 4.2. *If* $\lambda > 0$, *then*

$$
(4.32) \qquad f_r(x, y) = \sum_{n=0}^{\infty} r^n a_n \left(\frac{n+\lambda}{\lambda} \right) C_n^\lambda(x) C_n^\lambda(y) \geq 0,
$$

$$
-1 \leq x, y \leq 1, \quad 0 \leq r < 1,
$$

if and only if

$$
(4.33) \qquad a_n = \int_{-1}^{1} \frac{C_n^\lambda(x)}{C_n^\lambda(1)} d\mu(x), \, d\mu(x) \geq 0.
$$

This theorem is contained in Bochner [2]. A similar theorem is true for Jacobi series, but no proof has been published.

The following limiting relations are classical:

$$
(4.34) \qquad \lim_{\lambda \to \infty} \frac{C_n^\lambda(x)}{C_n^\lambda(1)} = x^n,
$$

$$
(4.35) \qquad \lim_{\lambda \to \infty} \lambda^{n/2} C_n^\lambda(x\lambda^{-1/2}) = H_n(x)/n!.
$$

If these are used in an appropriate way on (4.32) and (4.33), then the following result is formally obtained.

THEOREM 4.3. *If* $\sum_{n=0}^{\infty} a_n^2 < \infty$ *and*

$$\sum_{n=0}^{\infty} a_n \frac{H_n(x)H_n(y)}{2^n n!} \geqq 0, \qquad -\infty < x, y < \infty,$$

then

$$a_n = \int_{-1}^{1} r^n \, d\mu(r), \quad d\mu(r) \geqq 0.$$

The above argument is not a proof and I do not see how to give a proof along those lines. But Theorem 4.3 is true and was proved by O. V. Sarmanov and Bratoeva [1].

For Laguerre polynomials the analogous theorem is the following.

THEOREM 4.4. *Let* $\alpha > -1$ *and assume that* $\sum_{n=0}^{\infty} a_n^2 < \infty$. *If*

$$\sum_{n=0}^{\infty} a_n \frac{L_n^{\alpha}(x)L_n^{\alpha}(y)n!}{\Gamma(n + \alpha + 1)} \geqq 0, \qquad 0 \leqq x, y < \infty,$$

then

$$a_n = \int_{0}^{1} r^n \, d\mu(r), \quad d\mu(r) \geqq 0$$

(see I. O. Sarmanov [1]).

Gegenbauer's formulas have limiting results for Bessel functions which are quite useful:

$$\frac{J_{\nu}((x^2 + y^2 - 2xy \cos \theta)^{1/2})}{(x^2 + y^2 - 2xy \cos \theta)^{\nu/2}} = 2^{\nu}\Gamma(\nu) \sum_{m=0}^{\infty} (m + \nu) \frac{J_{m+\nu}(x)}{x^{\nu}} \frac{J_{m+\nu}(y)}{y^{\nu}} C_m^{\nu}(\cos \theta),$$

(4.36) $\nu \neq 0, -1, \cdots$.

For $\nu = 0$ the usual limiting argument gives

(4.37) $J_0((x^2 + y^2 - 2xy \cos \theta)^{1/2}) = J_0(x)J_0(y) + 2 \sum_{m=1}^{\infty} J_m(x)J_m(y) \cos m\theta.$

Integration using the orthogonality of $C_n^{\nu}(x)$ gives

(4.38) $\dfrac{J_{\nu}(x)}{x^{\nu}} \dfrac{J_{\nu}(y)}{y^{\nu}} = \dfrac{1}{2^{\nu}\Gamma(\nu + \frac{1}{2})\Gamma(\frac{1}{2})} \displaystyle\int_{0}^{\pi} \dfrac{J_{\nu}((x^2 + y^2 - 2xy \cos \theta)^{1/2})}{(x^2 + y^2 - 2xy \cos \theta)^{\nu/2}} \sin^{2\nu} \theta \, d\theta,$

$$\nu > -\tfrac{1}{2},$$

(see Watson [2]). Formula (4.38) has been used to set up a generalized translation operator for Hankel transforms (see Hirschman [1] and Cholewinski-Haimo [1]). Koornwinder [5] has given a summary of extensions of these formulas in the direction of replacing one of the J_{ν}'s by a J_{μ}. These are connected with the change of variables mentioned after (4.24). Watson [2] gives many other extensions, especially in the direction of replacing some of the J_{ν}'s by arbitrary cylinder functions $C_{\nu}(x) = AJ_{\nu}(x) + BY_{\nu}(x)$, where $Y_{\nu}(x)$ is the standard second solution to the Bessel differential equation. It seems likely that similar extensions can be made to the Šapiro-Koornwinder addition formula, but this has not been worked out yet.

There is another way of stating (4.38) which is usually attributed to Sonine [1]:

$$(4.39) \qquad \int_0^\infty J_\nu(at)J_\nu(bt)J_\nu(ct)t^{1-\nu}\,dt = \frac{2^{2\nu-1}\Delta^{2\nu-1}}{(abc)^\nu\Gamma(\nu+\tfrac12)\Gamma(\tfrac12)}, \qquad \nu > -\tfrac12,$$

when a, b, c are the sides of a triangle of area Δ; if a, b, c are not sides of a triangle the integral is zero (see Watson [2, 13.46 (3)]). Szegö [4] gives refinements when the triangle degenerates. Formulas (4.38) and (4.39) are inverses of each other as Hankel transforms.

A different extension of $\cos^2 x + \sin^2 x = 1$ to Bessel functions is Nicholson's integral:

$$(4.40) \qquad J_\nu^2(x) + Y_\nu^2(x) = \frac{4}{\pi^2}\int_0^\infty K_0(2x\sinh t)\cosh 2\nu t\,dt, \qquad \text{Re}\,(x) > 0,$$

(see Watson [2, 13.73 (1)]). Here $K_0(x)$ is yet another Bessel function. One way of defining it is by the integral

$$K_0(x) = \int_0^\infty e^{-x\cosh t}\,dt.$$

When $\nu = \pm\tfrac12$, (4.40) reduces to $\cos^2 x + \sin^2 x = 1$. The left-hand side is really

$$[J_\nu(x) + iY_\nu(x)][J_\nu(x) - iY_\nu(x)]$$

so Nicholson's integral is closely related to analogues of (4.38). Durand [1] has announced an extension of Nicholson's integral to Legendre functions. When $\nu = 0$ a different extension was given by Glasser [1]. Part of the interest in Nicholson's integral is the fact that it leads to many interesting inequalities for Bessel functions (see Lorch, Muldoon and Szego [1] and references given there to their earlier work). A nonformula proof of some of the consequences of Nicholson's integral is given by Hartman [1], [2]. The interest in his proof is that it works for equations other than the Bessel equation.

Finally there is the case of Laguerre polynomials. The ordinary Laguerre polynomials $L_n(x) = L_n^0(x)$ have an addition formula which was stated by Bateman [3, p. 457]:

$$(4.41)\quad \exp(xye^{i\theta})L_n(x^2 + y^2 - 2xy\cos\theta) = \sum_{k=0}^\infty (xye^{i\theta})^{k-n}\frac{n!}{k!}L_n^{(k-n)}(x^2)L_n^{(k-n)}(y^2).$$

A simple proof has been given by Carlitz [2]. This can be integrated to give

$$(4.42)\quad L_n(x^2)L_n(y^2) = \frac{1}{\pi}\int_0^\pi e^{xy\cos\theta}\cos(xy\sin\theta)L_n(x^2 + y^2 - 2xy\cos\theta)\,d\theta.$$

Watson [1] has generalized this to

$$
\frac{L_n^\alpha(x^2)L_n^\alpha(y^2)}{L_n^\alpha(0)} = \frac{2^{\alpha-1/2}\Gamma(\alpha+1)}{\sqrt{\pi}} \int_0^\pi e^{-xy\cos\theta} \frac{J_{\alpha-1/2}(xy\sin\theta)}{(xy\sin\theta)^{\alpha-1/2}}
$$

(4.43)

$$
\cdot L_n^\alpha(x^2+y^2+2xy\cos\theta)(\sin\theta)^{2\alpha}\,d\theta, \qquad \alpha > -\tfrac{1}{2}.
$$

This formula can be used to set up a generalized translation operator for Laguerre series but this operation is no longer a positive operator. It cannot be positive because of Theorem 4.4.

Three questions are raised by these formulas. First, find out what happens to (4.43) when $\alpha \to -\tfrac{1}{2}$. This is not hard, and the resulting formula was found by J. Boursma a number of years ago. As far as I know it is still unpublished. Next, find the extension of the addition formula (4.41) which has (4.43) as a consequence. This has recently been done by Koornwinder. And last, what is the basic underlying reason for these results? Peetre [1] has found this. The Laguerre functions $e^{-x/2}L_n(x)$ play the same role with respect to the Weyl transform and the Heisenberg group that Bessel functions play with respect to the ordinary Fourier transform on R^n. In particular $e^{-x/2}L_n(x)$ is the analogue of $J_0(x)$ in R^2. Similar results hold for $e^{-x/2}L_n^k(x)$ with respect to higher-dimensional analogues of the Heisenberg group. This suggests the problem of doing spherical harmonics in this setting and obtaining a direct proof of Koornwinder's addition formula. And it gives me the chance of stating the most interesting question in this area. We have a large class of special functions and many of them are now seen to arise in a very natural geometric or algebraic setting. How does one find the right spaces on which these functions live? This problem has been solved for $P_n^{(\alpha,\beta)}(x)$, $e^{-x/2}L_n^\alpha(x)$, $H_n(x)$, $J_\alpha(x)$, $K_n(x;p,N)$ and $\varphi_\lambda^{(\alpha,\beta)}(t)$, but not for most of the other functions mentioned in Lecture 2. Even when geometric or algebraic settings have been found they often only hold for restricted values of the parameters. For $P_n^{(\alpha,\beta)}(x)$ one natural restriction is $2\alpha, 2\beta$ integers (see Koornwinder [4]). For $\varphi_\lambda^{(\alpha,\beta)}(t)$ Flensted-Jensen [2] has partially removed these restrictions. As remarked above, Peetre [1] solved this question for $e^{-x/2}L_n^k(x)$, $k = 0, 1, \cdots$. For Hermite polynomials, $H_n(x)$, Orihara [1] has shown that an infinite-dimensional manifold which is a limit of spheres is a supporting set. It is likely that $L_n^\alpha(x)$, $\alpha = -\tfrac{1}{2}, 0, 1$, come from similar manifolds associated with projective spaces, but this has not been established. It would be more interesting to obtain $L_n^\alpha(x)$, 2α an integer, on an infinite-dimensional manifold in the same way Zernike and Brinkman [1] obtained $P_n^{(\alpha,\beta)}(x)$, $2\alpha, 2\beta$ integers (see Braaksma–Meulenbeld [1] and Koornwinder [4] as well).

Finally Krawtchouk polynomials come from a finite group when $p = k/(k+1)$, $k = 1, 2, \cdots$, (cf. Vere-Jones [1] for $p = \tfrac{1}{2}$, Dunkl and Ramirez [1] for the other cases) and from a more general algebraic structure called a hypergroup by Dunkl and Ramirez [1] for $\tfrac{1}{2} < p < 1$, $p \neq k/(k+1)$. There are similar results for $0 < p < \tfrac{1}{2}$, which Dunkl and Ramirez obtained by symmetry. When $p = \tfrac{1}{2}$ Dunkl announced a complete addition formula at this conference. This is the first addition formula not connected with a continuous group of transformations.

Many related but slightly different algebraic structures were considered by Miller [1]. The algebraic methods he has developed are very useful when deriving generating functions, but they have not been as successful in deriving addition formulas of the type given above. There are other addition formulas, for example, Graf's formula for Bessel functions, and analogues of this formula are not too hard to derive by Miller's methods. These are useful addition formulas, but not for the problems considered in these lectures.

The addition formulas given in this lecture are just special realizations of much more general addition formulas which have been used by Folland [1] to find the Poisson-Szegö kernel for the unit ball in complex n-space and by Goethals and Seidel in some as yet unpublished combinatorial work. This formula was also found by Šapiro and Koornwinder.

A very nice historical treatment of addition formulas is given in Koornwinder [10]. It should be read by anyone who wants to learn about these important results.

Added in proof. Dunkl [2] has now found an addition formula for all the Krawtchouk polynomials. The combinatorial work referred to above appears in Delsarte, Goethals and Seidel [1]. Extensions of Theorems 4.3 and 4.4 to general orthogonal polynomials on unbounded sets are given in Tyan and Thomas [1]. In particular a theorem like Theorem 4.4 holds for Charlier and Meixner polynomials.

LECTURE 5

Linearization of Products

One consequence of the addition formula for cosines is

(5.1) $$\cos m\theta \cos n\theta = \tfrac{1}{2}\cos(n+m)\theta + \tfrac{1}{2}\cos(n-m)\theta.$$

Since $\cos n\theta$ is a polynomial of degree n in $\cos\theta$ a natural generalization of (5.1) is the problem of determining, or at least saying something useful, about the coefficients in

(5.2) $$p_m(x)p_n(x) = \sum_{k=0}^{m+n} a(k,m,n)p_k(x).$$

The most important special case of (5.2) is

(5.3) $$x^m x^n = x^{m+n}.$$

Fortunately there are other classical results which are a bit more complicated, and so will tell more about the general type of result that might hold for wide classes of polynomials. For example, there is

(5.4) $$\frac{\sin(n+1)\theta}{\sin\theta}\frac{\sin(m+1)\theta}{\sin\theta} = \sum_{k=0}^{\min(m,n)}\frac{\sin(n+m+1-2k)\theta}{\sin\theta}$$

and the result of Ferrers [1, Example 10, p. 156] and Adams [1]:

(5.5) $$P_m(x)P_n(x) = \sum_{k=0}^{\min(m,n)}\frac{m+n+\tfrac{1}{2}-2k}{m+n+\tfrac{1}{2}-k}\frac{A_k A_{m-k}A_{n-k}}{A_{m+n-k}}P_{m+n-2k}(x),$$

where

(5.6) $$A_k = \frac{(\tfrac{1}{2})_k}{k!}.$$

In both of these cases, as in (5.1) and (5.3), the linearization coefficients are non-negative. In the case of ultraspherical polynomials an explicit formula is known for these coefficients and the formula is simple enough so that the sign properties are obvious. Explicitly it is

(5.7)
$$\begin{aligned}
C_m^\nu(x)C_n^\nu(x) = {} & \sum_{k=0}^{\min(m,n)}\frac{(m+n+\nu-2k)}{(m+n+\nu-k)}\frac{(\nu)_k(\nu)_{m-k}(\nu)_{n-k}(2\nu)_{m+n-k}}{k!(m-k)!(n-k)!(\nu)_{m+n-k}} \\
& \cdot\frac{(m+n-2k)!}{(2\nu)_{m+n-2k}}C_{m+n-2k}^\nu(x).
\end{aligned}$$

39

Dougall [2] first stated (5.7) but did not give his proof, which he said was complicated. Hsü [1] proved (5.7) by induction. Adams [1] remarked that both he and Ferrers found (5.5) by finding the coefficients for small values of m (Adams derived these coefficients for $m = 1, 2, 3, 4$ in his paper), guessing what the result would be for arbitrary m and n, and proving it by induction. This is not a bad way to discover the formula, but a more systematic method should be found. Neumann [1] found such a method, and it is still the most powerful method which is known. He computed a fourth order differential equation satisfied by $P_n(x)P_m(x)$ and used it to set up a recurrence relation for the linearization coefficients. Then he solved this recurrence relation. His argument is given in Hobson [1]. Hylleraas [1] set up a similar equation satisfied by $C_n^\nu(x)C_m^\nu(x)$, this time a fifth order equation, and obtained Dougall's formula (5.7) in a very natural way. Carlitz [4] proved (5.7) by another natural method. One of the results he used was Gegenbauer's addition formula for Bessel functions (4.36). This is interesting, for Dougall [2] claimed that it was possible to obtain Sonine's integral (4.39) from (5.7) and Hsü [1] gave the details. It is not hard to go from (4.39) to Gegenbauer's addition formula (4.36), so in some sense these results are equivalent. Carlitz's argument is one of the few I know in which one goes from Bessel functions to ultraspherical polynomials, rather than the other way around. One other instance was given in Lecture 3. Vilenkin [2, Chap. 3, § 8] gave a group theoretic proof of (5.5) but his proof of (5.7) was by induction. It should be possible to derive (5.7) directly when 2ν is an integer by the method Vilenkin used in the case $\nu = \frac{1}{2}$. Bailey [1] gives a proof of (5.5) by means of Whipple's transformation of a Saalschützian $_4F_3$ to a well-poised $_7F_6$. A. Verma called this reference to my attention. Bailey used the same method to generalize (5.5) to Legendre functions of both the first and the second kind. His results for functions of the second kind are interesting, but for functions of the first kind he does not seem to realize that Dougall's formula (4.7) contains his result. In particular, his comment on page 177 that his formula (3.3) does not generalize to noninteger values of his parameters is misleading, since Dougall's formula (5.7) provides a partial extension. This is just an instance of the fact that ultraspherical polynomials provide a nicer extension of Legendre polynomials than Legendre functions do for many problems. Dougall [3] gave yet another proof of (5.5). It is quite interesting.

For Jacobi polynomials the situation was far from satisfactory (we have seen this before). Hylleraas [1] had set up a recurrence relation for the coefficients in

$$(5.8) \qquad P_n^{(\alpha,\beta)}(x)P_m^{(\alpha,\beta)}(x) = \sum_{k=|n-m|}^{n+m} C(k, m, n)P_k^{(\alpha,\beta)}(x),$$

and he was able to solve this equation not only when $\alpha = \beta$ but when $\alpha = \beta + 1$. Again the coefficients were products of gamma functions. In the general case the most compact formula which has been obtained is as a double sum, Miller [2], and this sum has proved to be too complicated to use. Gasper [2], [3] used the recurrence relation of Hylleraas to prove the following theorem.

THEOREM 5.1. *The coefficients in* (5.8) *are nonnegative for all* k, m, n *if and only if* $\alpha \geq \beta > -1$ *and*

$$(\alpha + \beta + 1)(\alpha + \beta + 4)^2(\alpha + \beta + 6) \geq (\alpha - \beta)^2[(\alpha + \beta + 1)^2 - 7(\alpha + \beta + 1) - 24].$$

In particular the nonnegativity of the coefficients holds for $\alpha \geq \beta > -1, \alpha + \beta \geq -1$.

Applications of these results are given in Hirschman [2], Davis–Hirschman [1], Askey–Wainger [1], [2], Askey–Gasper [3], and Gasper [3].

For orthogonal polynomials another way of stating the problem of the signs of the coefficients in (5.2) is to consider the sign behavior of

(5.9)
$$\int_E p_k(x)p_m(x)p_n(x) \, d\alpha(x).$$

Dougall's formula (5.7) can be written as

(5.10)
$$\int_{-1}^1 C_k^\nu(x)C_m^\nu(x)C_n^\nu(x)(1 - x^2)^{\nu - 1/2} \, dx$$
$$= \frac{\alpha_{s-k}\alpha_{s-m}\alpha_{s-n}}{\alpha_s} \int_{-1}^1 [C_s^\nu(x)]^2(1 - x^2)^{\nu - 1/2} \, dx,$$

when $k + m + n = 2s$ is even and a triangle with sides k, m, n exists, and the integral is zero in all other cases. The numbers α_n are defined by

$$\alpha_n = \frac{(\nu)_n}{n!}.$$

In addition to the fact that (5.10) will be used in the next lecture, it is sometimes useful to consider (5.9) instead of (5.2). This arises when polynomials on a finite set of points are considered. For example, consider the Krawtchouk polynomials

$$K_n(x; p, N) = \sum_{k=0}^n \frac{(-n)_k(-x)_k}{(-N)_k k!} \left(\frac{1}{p}\right)^k, \qquad 0 < p < 1, \quad n = 0, 1, \cdots, N.$$

These polynomials are orthogonal on $\{0, 1, \cdots, N\}$ with respect to the binomial distribution $\binom{N}{x} p^x(1 - p)^{N-x}$. The right problem to consider for them is

(5.11)
$$\sum_{x=0}^N K_i(x; p, N)K_m(x; p, N)K_n(x; p, N) \binom{N}{x} p^x(1 - p)^{N-x} \geq 0.$$

Eagleson [1] proved (5.11) for $\frac{1}{2} \leq p < 1$ by using the generating function

(5.12)
$$(1 - pr)^{N-x}(1 + r - pr)^x = \sum_{k=0}^N K_k(x; p, N) \frac{(-N)_k p^k}{k!} r^k.$$

For $0 < p < \frac{1}{2}$ he obtained a similar result by considering

$$\mathcal{K}_n(x; p, N) = K_n(x; p, N)/K_n(N; p, N).$$

The inequality (5.11) can also be stated as

$$K_m(x;p,N)K_n(x;p,N) = \sum_{j=0}^{N} a(j,m,n;p,N)K_j(x;p,N),$$

(5.13)

$$m,n,x = 0,1,\cdots,N,$$

with

(5.14) $a(j,m,n;p,N) \geq 0,$ $\tfrac{1}{2} \leq p < 1,$

(5.15) $(-1)^{j+m+n}a(j,m,n;p,N) \geq 0,$ $0 < p \leq \tfrac{1}{2}.$

Observe that when $m + n > N$, (5.13) does not give the equality of two polynomials for all x, but only for $x = 0,1,\cdots,N$. This is why (5.11) is a more natural result than (5.13). A similar generating function argument can be used to show that

(5.16) $\displaystyle\int_{-\infty}^{\infty} H_k(x)H_m(x)H_n(x)\,e^{-x^2}\,dx \geq 0,$

(5.17) $(-1)^{k+m+n}\displaystyle\int_0^{\infty} L_k^\alpha(x)L_m^\alpha(x)L_n^\alpha(x)x^\alpha\,e^{-x}\,dx \geq 0,$ $\alpha > -1,$

(5.18) $(-1)^{k+m+n}\displaystyle\sum_{x=0}^{\infty} c_k(x;a)c_m(x;a)c_n(x;a)a^x/x! \geq 0,$ $a > 0,$

(5.19) $(-1)^{k+m+n}\displaystyle\sum_{k=0}^{\infty} M_k(x;\beta,c)M_m(x;\beta,c)M_n(x;\beta,c)(\beta)_x c^x/x! \geq 0,$

$$0 < c < 1, \quad \beta > 0.$$

And since all of these discrete polynomials are self-dual, i.e.,

$$p_n(x) = p_x(n), \qquad\qquad x,n = 0,1,\cdots,$$

each of the results (5.11), (5.18), (5.19) gives a linearization result of the type considered in the last lecture. Eagleson [1] gave an application of his result to find the values of r for which the kernel (2.51) for Krawtchouk polynomials is nonnegative. When $\tfrac{1}{2} \leq p < 1$ it is nonnegative for $1 - 1/p \leq r \leq 1$. Recall that we stated a similar theorem for the Poisson kernel for Jacobi series in Lecture 2.

The Hermite result can be stated as

$$H_m(x)H_n(x) = \sum_{k=0}^{\min(m,n)} 2^k k!\binom{m}{k}\binom{n}{k}H_{m+n-2k}(x).$$

This formula was used by Ginibre [1]. He only needed the nonnegativity of the coefficients.

Consider boxes A, B, and C which hold k, m, and n objects respectively. The objects in A are labeled $a_1,\cdots,a_k$; in B, $b_1,\cdots,b_m$; and in C, $c_1,\cdots,c_n$. How many ways can these objects be moved so that no element is in the box it started in and each box has as many elements after the moves as before? We distinguish between the objects $a_1,\cdots,a_k$ but not the places they hold in the box they move to. For example, if $k = 2$, $m = 2$, $n = 1$, the two cases $b_1,c_1;a_1,a_2;b_2$ and $b_1,c_1;$ $a_2,a_1;b_2$ are not distinguished; but the two cases $b_1,c_1;a_1,a_2;b_2$ and $b_2,c_1;$ $a_1,a_2;b_1$ are treated as different.

The answer to this problem is

$$(-1)^{k+m+n} \int_0^\infty L_k(x)L_m(x)L_n(x)\, e^{-x}\, dx.$$

This was pointed out by J. Gillis after these lectures had been given. G. Andrews showed me how to derive this using MacMahon's master theorem. Much more is true. The integrals

(5.20)
$$(-1)^{n_1+\cdots+n_N} \int_0^\infty L_{n_1}(x) \cdots L_{n_N}(x)\, e^{-x}\, dx$$

give the number of derangements (or displacements) of $n_1 + \cdots + n_N$ objects when there are n_1 of the first type, n_2 of the second, $\cdots$, n_N of the Nth type. For the generating function of the integrals (5.20) and of the derangements is

$$\frac{1}{1 - p_2 - 2p_3 - \cdots - (N-1)p_N},$$

where p_i are the elementary symmetric functions defined by

$$(x - r_1) \cdots (x - r_N) = x^N - p_1 x^{N-1} + p_2 x^{N-2} \cdots + (-1)^N p_N.$$

(see MacMahon [1, § 68] for a derivation of the combinatorial generating function). The orthogonal polynomial generating function can easily be derived by use of

$$\frac{e^{-xr/(1-r)}}{1-r} = \sum_{n=0}^\infty L_n(x)r^n.$$

J. Gillis and S. Even independently found a proof of this formula using generating functions and recurrence relations. The classical problem of derangements of n different objects is easily solved by this formula, for it is

$$(-1)^n \int_0^\infty [L_1(x)]^n\, e^{-x}\, dx = \int_0^\infty (x-1)^n\, e^{-x}\, dx = I_n$$

and the integral can easily be evaluated by integration by parts. For

$$\frac{I_n}{n!} = \frac{(-1)^n}{n!} + \frac{I_{n-1}}{(n-1)!} \quad \text{and} \quad I_1 = 0$$

so

$$I_n = n! \sum_{k=0}^n \frac{(-1)^k}{k!}.$$

See Gillis-Weiss [1] for recurrence relations connecting adjacent integrals of the form (5.17), and MacMahon [1, § III, Chaps. II, III and IV] for other combinatorial results. One of them is equivalent to

$$(-1)^{3k} \int_0^\infty [L_k(x)]^3\, e^{-x}\, dx = 1 + \binom{k}{1}^3 + \binom{k}{2}^3 + \cdots + \binom{k}{k}^3,$$

a result which I find far from obvious.

It seems likely that other interesting combinatorial connections can be developed for the corresponding sum (5.19) of Meixner polynomials.

So far we have treated the classical orthogonal polynomials as hypergeometric functions. Nowhere have we really used orthogonality except in the most elementary way. Now we shall treat orthogonal polynomials directly, and thus the results in the rest of this lecture are not restricted to the classical polynomials.

Polynomials orthogonal on an infinite set satisfy

$$(5.21) \qquad\qquad xp_n(x) = p_{n+1}(x) + a_n'p_n(x) + b_np_{n-1}(x), \qquad\qquad n \geq 1,$$

$p_0(x) = 1$, $p_1(x) = x + c$, c and a_n' real and $b_n > 0$, $n = 1, 2, \cdots$. Favard [1] (see also Atkinson [1], Freud [1]) proved the converse, that a set of polynomials which satisfies (5.21) is orthogonal with respect to a positive measure. This measure is not necessarily unique because a moment problem does not necessarily have a unique solution. The measure has bounded support if and only if a_n' and b_n are bounded, and in this case the measure is uniquely determined. When the polynomials are orthogonal on a bounded set, if the linearization coefficients are nonnegative, and if a slight boundedness condition is satisfied, then there is a natural Banach algebra associated with the polynomials which is analogous to l^1 of the circle (see Askey [4], Askey–Wainger [2], Askey–Gasper [1], Gasper [3] and Igari–Uno [1] for examples).

To connect the recurrence relation (5.21) with the problem we have treated add $cp_n(x)$ to both sides and rewrite the recurrence relation as

$$(5.22) \qquad\qquad p_1(x)p_n(x) = p_{n+1}(x) + a_np_n(x) + b_np_{n-1}(x).$$

This is just the case $m = 1$ of (5.2). We shall prove the following theorem (see Askey [4]).

THEOREM 5.2. *Let $p_n(x)$ satisfy (5.22), $n = 1, 2, \cdots$, and define the linearization coefficients by*

$$(5.23) \qquad\qquad p_m(x)p_n(x) = \sum_{k=|n-n|}^{n+m} a(k, m, n)p_k(x).$$

If $a_n \geq 0$, $b_n \geq 0$, $a_{n+1} \geq a_n$ and $b_{n+1} \geq b_n$, $n = 1, 2, \cdots$, then $a(k, m, n) \geq 0$.

Proof. By symmetry assume $m < n$ and that $a(k, j, l) \geq 0$, $j = 0, 1, \cdots, m$, $j < l$. Then

$$p_{m+1}(x)p_n(x) = [p_1(x)p_m(x) - a_mp_m(x) - b_mp_{m-1}(x)]p_n(x)$$

$$= p_m(x)[p_{n+1}(x) + a_np_n(x) + b_np_{n-1}(x)]$$

$$- a_mp_m(x)p_n(x) - b_mp_{m-1}(x)p_n(x)$$

$$= p_m(x)p_{n+1}(x) + (a_n - a_m)p_m(x)p_n(x)$$

$$+ (b_n - b_m)p_{m-1}(x)p_n(x) + b_n[p_m(x)p_{n-1}(x) - p_{m-1}(x)p_n(x)].$$

The first three terms can be written as sums of $p_k(x)$ times nonnegative coefficients by the induction assumption and the monotonicity of a_k and b_k. Also $b_n \geq 0$ so it is sufficient to prove that

$$p_m(x)p_{n-1}(x) - p_{m-1}(x)p_n(x) = \sum_{j=0}^{m+n-2} c(j)p_j(x)$$

with $c(j) \geq 0$. This follows from another induction. For

$$p_{m+1}(x)p_n(x) - p_m(x)p_{n+1}(x) = \sum_{j=0}^{m+n} d(j)p_j(x) + p_m(x)p_{n-1}(x) - p_{m-1}(x)p_n(x)$$

$$= \sum_{j=0}^{m+n} e(j)p_j(x) + p_1(x)p_{n-m}(x) - p_{n-m+1}(x)$$

$$= \sum_{j=1}^{m+n} e(j)p_j(x) + a_{n-m}p_{n-m}(x) + b_{n-m}p_{n-m-1}(x),$$

and all these coefficients are nonnegative by induction and assumptions.

It is possible to state the theorem in a different way so that it is seen to be a discrete analogue of Weinberger's maximum theorem for hyperbolic equations (see Askey [3]).

Surprisingly this simple argument gives a theorem which is strong enough to include most of the examples given above. The two results which are not contained in Theorem 5.2 are Eagleson's result for Krawtchouk polynomials and Gasper's result for Jacobi polynomials when α and β are small. When $\alpha \geq \beta$ and $\alpha + \beta \geq 1$ the assumptions of Theorem 5.1 are satisfied. It is likely that there is a stronger theorem than this one which will give the same conclusion for $\alpha \geq \beta \geq -\frac{1}{2}$. Recall that Koornwinder was able to modify Weinberger's argument to obtain the dual result under these conditions. However, I have been unable to find such a theorem. Part of the problem comes from the argument, which proves too much. It shows that certain coefficients are positive because they have increased from something which was nonnegative. And this is too strong a conclusion.

There are other interesting sets of polynomials orthogonal on a bounded set to which Theorem 5.2 applies. The most interesting is the set of Pollaczek polynomials (see Szegö [9, Appendix] and references given there). These polynomials satisfy a recurrence relation of the form (5.21) with a_n' and b_n rational functions. If the polynomials defined by (5.21) are orthogonal on $[-1, 1]$ and the weight function is positive there, then $\lim_{n \to \infty} a_n = 0$ and $\lim_{n \to \infty} b_n = \frac{1}{4}$ (see Szegö [9, Chap. 12]). In the case of Jacobi polynomials $a_n = O(1/n^2)$ and $b_n = \frac{1}{4} + O(1/n^2)$, while in the case of the Pollaczek polynomials which are not Jacobi polynomials, $a_n = c/n + O(1/n^2)$ and $b_n = \frac{1}{4} + d/n + O(1/n^2)$ with either $c \neq 0$ or $d \neq 0$. We can ask if this is always true in the following sense. In the case of Jacobi polynomials,

$$\int_{-1}^{1} \frac{\log w(x)\,dx}{\sqrt{1-x^2}} > -\infty$$

while for Pollaczek polynomials which are not Jacobi polynomials, the weight function vanishes so rapidly at $x = 1$ that

$$\int_{-1}^{1} \frac{\log w(x)\, dx}{\sqrt{1 - x^2}} = -\infty.$$

One additional reason for thinking this is true is that polynomials grow rapidly at points where the weight function vanishes. And this growth is reflected in the recurrence relation. The difference may come from $\sum n^{-2} < \infty$ and $\sum n^{-1} = +\infty$.

Interesting integral representations have been found for the Pollaczek polynomials which are close to Legendre polynomials but not for the other polynomials. These should be found and the asymptotics worked out. Also it is likely that Pollaczek polynomials provide a solution to a problem of Turán. This is to find a weight function $w(x) > 0$, $-1 < x < 1$, $w(x) = 0$, $|x| > 1$ for which the Lagrange interpolation polynomials at the zeros of the corresponding orthogonal polynomials diverge in L^p for all $p > 2$ for some continuous function. For each fixed $p > 2$ there is a value of α for which convergence fails when $w(x) = (1 - x)^\alpha$, and as $p \to 2$, α must increase to infinity (see Askey [8]). This strongly suggests that any of the Pollaczek weight functions will give an example. There are many other problems which these polynomials suggest. Novikoff has never published his thesis [1], but a summary exists in the appendix to Szegö [9], and a copy can be obtained from University Microfilms. It is a good place to start.

LECTURE 6

Rational Functions with Positive Power Series Coefficients

Turán once remarked to me that special functions were misnamed. They should be called useful functions. A striking illustration of this comes when considering an old problem of Friedrichs and Lewy, which arose from a finite difference approximation to the wave equation in two space. They conjectured that

$$(6.1) \quad \frac{1}{(1 - r)(1 - s) + (1 - r)(1 - t) + (1 - s)(1 - t)} = \sum_{k,m,n=0}^{\infty} a(k, m, n) r^k s^m t^n$$

has positive power series coefficients. Szegö [4] gave a proof using Sonine's integral of the product of three Bessel functions (4.39). He also extended the original problem of Friedrichs and Lewy in two directions, to more variables and to different powers, and solved these problems. Lastly he translated this problem into an equivalent problem about the positivity of integrals of products of Laguerre polynomials. This will be our starting point, since not only can we use results of the last lecture to prove Szegö's theorems but stronger positivity theorems are suggested by this formulation and proof.

Laguerre polynomials $L_n^\alpha(x)$ have the generating function

$$(6.2) \quad (1 - r)^{-\alpha - 1} \exp(-xr/(1 - r)) = \sum_{n=0}^{\infty} L_n^\alpha(x) r^n.$$

Recall the orthogonality relation

$$(2.43a) \quad \int_0^\infty L_m^\alpha(x) L_n^\alpha(x) x^\alpha e^{-x} dx = \frac{\Gamma(n + \alpha + 1)}{n!} \delta_{m,n}.$$

To translate (6.1) into a more manageable problem observe that

$$(6.3) \quad \begin{aligned} &\frac{1}{(1 - r)(1 - s) + (1 - r)(1 - t) + (1 - s)(1 - t)} \\ &= \frac{1}{(1 - r)(1 - s)(1 - t)} \cdot \frac{1}{1/(1 - r) + 1/(1 - s) + 1/(1 - t)} \\ &= \int_0^\infty \frac{e^{-x/(1-r)}}{1 - r} \cdot \frac{e^{-x/(1-s)}}{1 - s} \cdot \frac{e^{-x/(1-t)}}{1 - t} dx. \end{aligned}$$

47

Thus a problem which was originally complicated because of the intermixing of the variables r, s and t has been changed into another problem with r, s and t separated. However, the problem has not disappeared; it is contained in the integration. The next step is clearly to expand in power series and integrate term by term. This cannot be done directly with $e^{-x/(1-r)}/(1-r)$ since the coefficient of r^n is a power series in x which is not easily recognizable. However, the simple reduction

$$\frac{e^{-x/(1-r)}}{1-r} = e^{-x} \cdot \frac{e^{-xr/(1-r)}}{1-r}$$

gives

(6.4)
$$a(k, m, n) = \int_0^\infty L_k(x) L_m(x) L_n(x) e^{-3x} \, dx.$$

A similar argument leads to

(6.5)
$$\frac{1}{[f'(1)]^{\alpha+1}} = \sum \frac{a^\alpha(k_1, \cdots, k_n)}{\Gamma(\alpha+1)} r_1^{k_1} \cdots r_n^{k_n}$$

with

(6.6)
$$a^\alpha(k_1, \cdots, k_n) = \int_0^\infty L_{k_1}^\alpha(x) \cdots L_{k_n}^\alpha(x) x^\alpha e^{-nx} \, dx$$

and

$$f(x) = (x - r_1) \cdots (x - r_n).$$

Szegö [4] had these formulas and (6.4) suggested Sonine's integral (4.39) to him. He then used (2.45W) as a substitute for the generating function (6.2) and proved the following theorem.

THEOREM 6.1. *If* $f(x) = (x - r_1) \cdots (x - r_n)$, *then* $[f'(1)]^{-\alpha-1}$ *is an absolutely monotonic function for* $\alpha \geq -\frac{1}{2}$, *i.e., the coefficients* $a^\alpha(k_1, \cdots, k_n)$ *in* (6.5) *are nonnegative. These coefficients are strictly positive when* $n > 4(\alpha+1)/(2\alpha+1)$, $\alpha > -\frac{1}{2}$, *and for* $n \geq 3$ *when* $\alpha \geq 0$.

We shall prove this theorem when $n = 3$ using Szegö's formula

(6.7)
$$a^\alpha(k, m, n) = \int_0^\infty L_k^\alpha(x) L_m^\alpha(x) L_n^\alpha(x) x^\alpha e^{-3x} \, dx.$$

One integral which we know to be nonnegative and somewhat similar to the integral in (6.7) is Dougall's integral (5.10). Clearly

(6.8)
$$\int_{-1}^1 P_k^{(\alpha,\alpha)}(x) P_m^{(\alpha,\alpha)}(x) P_n^{(\alpha,\alpha)}(x)(1 - x^2)^\alpha \, dx \geq 0, \qquad \alpha \geq -\frac{1}{2}.$$

To use (6.8) there must be a connection between $P_n^{(\alpha,\alpha)}(x)$ and $L_n^\alpha(x)$. This connection is provided by

(6.9)
$$c_n \left(\frac{1-x}{2} \right) \frac{P_n^{(\alpha,\beta+1)}(x)}{P_n^{(\beta+1,\alpha)}(1)} = \frac{P_n^{(\alpha,\beta)}(x)}{P_n^{(\beta,\alpha)}(1)} + \frac{P_{n+1}^{(\alpha,\beta)}(x)}{P_{n+1}^{(\beta,\alpha)}(1)}$$

with

$$c_n = \frac{2n + \alpha + \beta + 2}{(\beta + 1)},$$

or as it is usually written,

(6.10) $$(1 + x)P_n^{(\alpha,\beta+1)}(x) = \frac{(2n + 2)}{(2n + \alpha + \beta + 2)} P_{n+1}^{(\alpha,\beta)}(x) + \frac{(2n + 2\beta + 2)}{(2n + \alpha + \beta + 2)} P_n^{(\alpha,\beta)}(x).$$

Also we need the limit relation

(6.11)
$$\lim_{\beta \to \infty} P_n^{(\alpha,\beta)}\left(1 - \frac{2x}{\beta} \right) = L_n^\alpha(x).$$

Observe that the coefficients of $P_n^{(\alpha,\beta)}(x)$ and $P_{n+1}^{(\alpha,\beta)}(x)$ in (6.10) are positive for $\alpha, \beta > -1$. Both of these formulas are quite general and easily motivated (see Askey–Gasper [3]). Together they lead immediately to the weak version of Theorem 6.1 that $a^\alpha(k, m, n) \geq 0, \alpha \geq -\frac{1}{2}$. For (6.10) and (6.8) together give

(6.12) $$\int_{-1}^{1} P_k^{(\alpha,\alpha+j)}(x)P_m^{(\alpha,\alpha+j)}(x)P_n^{(\alpha,\alpha+j)}(x)(1-x)^\alpha(1+x)^{\alpha+3j} \, dx \geq 0, \quad \alpha \geq -\tfrac{1}{2},$$

and then (6.11) gives

(6.13)
$$\int_0^\infty L_k^\alpha(x)L_m^\alpha(x)L_n^\alpha(x)x^\alpha \, e^{-3x} \, dx \geq 0, \qquad\qquad \alpha \geq -\tfrac{1}{2}.$$

The details are given in Askey–Gasper [3].

The strict positivity for $\alpha \geq 0$ follows from

$$\frac{1}{[\cdot]^{\alpha+1}} = \frac{1}{[\cdot]^{(\alpha-1)/2+1}} \cdot \frac{1}{[\cdot]^{(\alpha-1)/2+1}}$$

so that

$$\left[\Gamma\left(\frac{\alpha+1}{2} \right) \right]^2 \int_0^\infty L_k^\alpha(x)L_m^\alpha(x)L_n^\alpha(x)x^\alpha \, e^{-3x} \, dx$$

$$= \Gamma(\alpha+1) \sum_{a=0}^{k} \sum_{b=0}^{m} \sum_{c=0}^{n} \int_0^\infty L_{k-a}^{(\alpha-1)/2}(x)L_{m-b}^{(\alpha-1)/2}(x)L_{n-c}^{(\alpha-1)/2}(x)x^{(\alpha-1)/2} \, e^{-3x} \, dx$$

$$\cdot \int_0^\infty L_a^{(\alpha-1)/2}(x)L_b^{(\alpha-1)/2}(x)L_c^{(\alpha-1)/2}(x)x^{(\alpha-1)/2} \, e^{-3x} \, dx.$$

When $\alpha \geq 0$ all the terms on the right are nonnegative, so it will be sufficient to find one positive term. The term with $a = k$, $b = m$, $c = 0$ is positive since

$$(6.14) \qquad \int_0^\infty e^{-\varepsilon x} L_n^\alpha(x) L_m^\alpha(x) x^\alpha e^{-x} \, dx > 0, \qquad \alpha > -1, \varepsilon > 0.$$

This integral is dual to the positive kernels considered in Lecture 2. Since orthogonal polynomials satisfy second order difference equations in n (the recurrence relation in n) and the classical polynomials of Jacobi, Laguerre and Hermite satisfy second order differential equations in x, it is natural to see what happens to a given formula when x and n are interchanged and integration and summation are substituted for each other. When this is done in

$$(6.15) \qquad \sum_{n=0}^\infty e^{-\varepsilon n} \frac{L_n^\alpha(x) L_n^\alpha(y)}{\int_0^\infty [L_n^\alpha(x)]^2 x^\alpha e^{-x} \, dx},$$

formula (6.14) is obtained. For when $\varepsilon = 0$, (6.14) is a constant multiple of the discrete delta function and (6.15) is the formal reproducing kernel; i.e., the delta function, when $\varepsilon = 0$. So the positivity of (6.14) is not surprising, and it can be proved either by explicitly computing (6.14) or from the differential difference equation it satisfies. For this proof, see Karlin–McGregor [1] or Szegö [9, Problems 81 and 82]. Karlin and McGregor also have given a probabilistic meaning to (6.14) (see Karlin–McGregor [1]).

Surprisingly, the dual result to Theorem 6.1 (when stated as the nonnegativity of (6.7)) is completely false. In fact, it is an easy consequence of Sarmanov's Theorem 4.4. For the dual problem is to find a sequence c_n for which

$$(6.16) \qquad \sum_{n=0}^\infty \frac{c_n L_n^\alpha(x) L_n^\alpha(y) L_n^\alpha(z)}{h_n^\alpha} \geq 0, \qquad 0 \leq x, y, z < \infty.$$

From Theorem 4.4,

$$(6.17) \qquad c_n L_n^\alpha(z) = \int_0^1 t^n \, d\mu_z(t), \; d\mu_z(t) \geq 0, \qquad 0 \leq z < \infty.$$

Letting $z = 0$ gives $c_n \geq 0$, and then for $n \geq 1$ there is a z so that $L_n^\alpha(z) < 0$. Then the left-hand side is nonpositive and the right-hand side is nonnegative, and so both are zero. Thus $c_n = 0$, $n = 1, 2, \cdots$, and the only way to have (6.16) hold is to have the left-hand side reduce to a constant. This is not the case with respect to Szegö's result, since

$$(6.18) \qquad \int_0^\infty e^{-2x} L_k^\alpha(x) L_m^\alpha(x) L_n^\alpha(x) x^\alpha e^{-x} \, dx \geq 0, \qquad \alpha \geq -\tfrac{1}{2}.$$

There are two natural ways of trying to extend the above proof to more than three variables. We can use

$$(6.19) \qquad \int_{-1}^1 \prod_{i=1}^n P_{k_i}^{(\alpha, \alpha)}(x)(1 - x^2)^\alpha \, dx \geq 0, \qquad \alpha \geq -\tfrac{1}{2}, \quad n = 0, 1, \cdots,$$

for any sequence of k_i, which follows from

$$(6.20) \qquad P_n^{(\alpha,\alpha)}(x)P_m^{(\alpha,\alpha)}(x) = \sum_{k=|n-m|}^{n+m} c_{k,m,n}^{\alpha} P_k^{(\alpha,\alpha)}(x),$$

$c_{k,m,n}^{\alpha} \geq 0$ when $\alpha \geq -\frac{1}{2}$. This proof goes through with no trouble (see Askey–Gasper [3]).

Or we can try to iterate directly using (6.18). Another way of stating (6.18) is

$$(6.21) \qquad e^{-2x}L_n^{\alpha}(x) e^{-2x}L_m^{\alpha}(x) = \sum_{k=0}^{\infty} b^{\alpha}(k,m,n) e^{-2x}L_k^{\alpha}(x),$$

with $b^{\alpha}(k,m,n) \geq 0$ when $\alpha \geq -\frac{1}{2}$. Then (6.21) can be iterated to obtain

$$(6.22) \qquad e^{-2x}L_j^{\alpha}(x) e^{-2x}L_n^{\alpha}(x) e^{-2x}L_m^{\alpha}(x) = \sum_{k=0}^{\infty} c^{\alpha}(k,j,m,n) e^{-2x}L_k^{\alpha}(x),$$

$c^{\alpha}(k,j,m,n) \geq 0$, $\alpha \geq -\frac{1}{2}$. These two methods lead to different results. The first method leads to

$$(6.23) \qquad \int_0^{\infty} L_j^{\alpha}(x)L_k^{\alpha}(x)L_m^{\alpha}(x)L_n^{\alpha}(x)x^{\alpha} e^{-4x} dx \geq 0, \qquad \alpha > -\frac{1}{2},$$

while the second only leads to

$$(6.24) \qquad \int_0^{\infty} L_j^{\alpha}(x)L_k^{\alpha}(x)L_m^{\alpha}(x)L_n^{\alpha}(x)x^{\alpha} e^{-5x} dx \geq 0, \qquad \alpha \geq -\frac{1}{2}.$$

The first is much stronger than the second for the positivity of (6.14) leads to the following theorem.

THEOREM 6.2. *If $\alpha > -1$ and*

$$a_n = \int_0^{\infty} f(x)L_n^{\alpha}(x)x^{\alpha} e^{-x} dx \geq 0, \qquad n = 0, 1, \cdots,$$

then

$$a_n(\varepsilon) = \int_0^{\infty} f(x) e^{-\varepsilon x}L_n^{\alpha}(x)x^{\alpha} e^{-x} dx > 0, \qquad n = 0, 1, \cdots, \qquad \varepsilon > 0,$$

unless $f(x) = 0$, $x \geq 0$.

It would be nice if a stronger result than (6.21) held, one which did not become weaker than it should under iteration. Such a theorem would follow if we could replace $e^{-2x}L_n^{\alpha}(x)$ by $e^{-x}L_n^{\alpha}(x)$.

For $\alpha = -\frac{1}{2}$ this not possible by a formula of Titchmarsh [1]. But it is for $\alpha = 0$. More generally, the following theorem holds (see Askey–Gasper [4]).

THEOREM 6.3. *If $\alpha \geq (-5 + \sqrt{17})/2$, then*

$$(6.25) \qquad \int_0^{\infty} L_k^{\alpha}(x)L_m^{\alpha}(x)L_n^{\alpha}(x)x^{\alpha} e^{-2x} dx \geq 0$$

and the only case of equality is $\alpha = (-5 + \sqrt{17})/2, k = m = n = 1$. *Also*

$$\int_0^\infty \prod_{i=1}^n L_{k_i}^\alpha(x) x^\alpha e^{-(n-1)x}\, dx \geq 0, \qquad \alpha \geq (-5 + \sqrt{17})/2.$$

The proof of Theorem 6.3 is quite interesting. Recall that (6.4) came from the rational function (6.1) and more generally (6.7) came from the expansion of

(6.26)
$$\frac{1}{[(1 - r)(1 - s) + (1 - r)(1 - t) + (1 - s)(1 - t)]^{\alpha + 1}}.$$

However, it does not seem to be possible to prove Theorem 6.1 directly from this function. Kaluza [1] gave an involved proof of Theorem 6.1 when $\alpha = 0, n = 3$ without using special functions, but so far his proof has not been extended to prove all of Theorem 6.1. With respect to (6.25) the opposite is true. I do not know how to prove (6.25) directly, but it can be translated back to a problem involving power series coefficients of an algebraic function,

(6.27) $[2 - (r + s + t) + rst]^{-\alpha - 1} = \displaystyle\sum_{k,m,n=0}^\infty r^k s^m t^n \frac{\int_0^\infty L_k^\alpha(x) L_m^\alpha(x) L_n^\alpha(x) x^\alpha e^{-2x}\, dx}{\Gamma(\alpha + 1)}.$

The left-hand side can also be written as

(6.28)
$$(3/2)^{+\alpha+1}\left[(1 - r)(1 - s)\left(1 + \frac{t}{2}\right) + (1 - r)\left(1 + \frac{s}{2}\right)(1 - t) \right.$$
$$\left. + \left(1 + \frac{r}{2}\right)(1 - s)(1 - t)\right]^{-\alpha - 1},$$

and the function (6.26) can be written as

(6.29)
$$\left[(1 - r)(1 - s)\left(1 + \frac{t}{\infty}\right) + (1 - r)\left(1 + \frac{s}{\infty}\right)(1 - t) \right.$$
$$\left. + \left(1 + \frac{r}{\infty}\right)(1 - s)(1 - t)\right]^{-\alpha - 1}.$$

The function on the left-hand side of (6.27) can be expanded in a power series using the multinomial theorem, and the coefficient of $r^k s^m t^n$ is a positive multiple of

(6.30) $_3F_2(-k, -m, -n; (-\alpha - k - m - n)/2, (1 - \alpha - k - m - n)/2; 1).$

When $\alpha = -\frac{1}{2}$ this function is Saalschützian, i.e., it is

$$_3F_2\left(\begin{matrix} a, b, c \\ d, e \end{matrix}; 1\right)$$

with $a + b + c + 1 = d + e$ and one of a, b, c is a negative integer, so the series terminates. Using Saalschütz's formula (Bailey [2]) this $_3F_2$ is a positive multiple of $(-1)^{k+m+n}\Gamma(m + n - k + \frac{1}{2})\Gamma(m + k - n + \frac{1}{2})\Gamma(n + k - m + \frac{1}{2})$ and this is not nonnegative for all k, m, n. This is equivalent to a formula of Titchmarsh [1]. For

$\alpha = 0, 1, \cdots$ the $_3F_2$ in (6.30) has not been summed but it is not very hard to prove it is positive. Assume $k \leqq m$, $k \leqq n$ and reverse the sum in (6.27) to obtain

$$\sum_{j=0}^{k} \frac{(-k)_j(-m)_j(-n)_j}{\left(\dfrac{-k-m-n-\alpha}{2}\right)_j \left(\dfrac{1-k-m-n-\alpha}{2}\right)_j j!}$$

$$= \frac{(-k)_k(-m)_k(-n)_k}{\left(\dfrac{-k-m-n-\alpha}{2}\right)_k \left(\dfrac{1-k-m-n-\alpha}{2}\right)_k k!}$$

(6.31)
$$\cdot \sum_{j=0}^{k} \frac{(-k)_j \left(\dfrac{\alpha+1+m+n-k}{2}\right)_j \left(\dfrac{\alpha+2+m+n-k}{2}\right)_j}{(m+1-k)_j(n+1-k)_j j!}$$

$$= \frac{(-1)^k m! \, n! \, 2^{2k} \Gamma(m+n+\alpha+1-k)}{(m-k)!(n-k)! \Gamma(m+n+k+\alpha+1)}$$

$$\cdot {}_3F_2\left(\begin{matrix} -k, (\alpha+1+m+n-k)/2, (\alpha+2+m+n-k)/2 \\ m+1-k, n+1-k \end{matrix} ; 1\right).$$

This changes a sum whose terms have five factors which change sign when going from one term to the next, to a series with only one factor which changes sign per term, but we have to show that $(-1)^k {}_3F_2$ is positive. This can be done when $\alpha = 0$ by use of the Kummer–Thomae–Whipple formula

(6.32) $\quad {}_3F_2\left(\begin{matrix} a, b, c \\ d, e \end{matrix} ; 1\right) = \dfrac{\Gamma(e)\Gamma(d+e-a-b-c)}{\Gamma(e-c)\Gamma(d+e-a-b)} {}_3F_2\left(\begin{matrix} d-a, d-b, c \\ d, d+e-a-b \end{matrix} ; 1\right).$

Watson's proof of (6.32) is particularly simple. He used

(6.33) $\qquad {}_2F_1(a, b; d; x) = (1-x)^{d-a-b} {}_2F_1(d-a, d-b; d; x).$

This formula of Euler [4] is one of the basic formulas for hypergeometric functions. For example, as Dougall [1] remarked, Saalschütz's formula comes from multiplying the two functions on the right and equating coefficients. Euler [4] proved (6.33) by changing variables in the differential equation satisfied by $_2F_1(a, b; d; x)$:

$$x(1-x)y'' + [d-(a+b+1)x]y' - aby = 0, \quad y = {}_2F_1(a, b; d; x).$$

Kummer [1] proved (6.33) by changing variables (let $t = 1 - y$) in the integral

(6.34) $\qquad {}_2F_1(a, b; d; x) = \dfrac{\Gamma(d)}{\Gamma(d-b)\Gamma(b)} \int_0^1 (1-xt)^{-a} t^{b-1}(1-t)^{d-b-1} \, dt$

and simplifying to get

(6.35) $\qquad {}_2F_1(a, b; d; x) = (1-x)^{-a} {}_2F_1(a, d-b; d; x/(x-1)).$

Then the symmetry of $_2F_1(a, b; d; x)$ in a and b was used, and (6.35) was used a second time to obtain (6.33). Finally (6.33) is multiplied by $x^{c-1}(1 - x)^{e-c-1}$ and integrated from $x = 0$ to $x = 1$, and formula (6.32) is obtained. This is one of a large number of transformation formulas which were systematically found by Thomae [1] and later by Whipple [1]. Whipple also organized these formulas into a manageable group. However (6.32) was first stated by Kummer [1, p. 172], and it is likely that the proof which was given above was the method Kummer used to discover this transformation formula.

Finally to complete the proof which was started above use (6.32) on (6.31) to get

$$(-1)^k {}_3F_2 \left(\begin{matrix} -k, (\alpha + 1 + m + n - k)/2, (\alpha + 2 + m + n - k)/2 \\ m + 1 - k, n + 1 - k \end{matrix} ; 1 \right)$$

$$= (-1)^k \frac{\Gamma(n + 1 - k)\Gamma(\tfrac{1}{2} - \alpha)}{\Gamma(n + 1)\Gamma(\tfrac{1}{2} - \alpha - k)}$$

(6.36)
$$\cdot {}_3F_2 \left(\begin{matrix} (m - k - n - \alpha)/2, (m - k - n - \alpha + 1)/2, -k \\ m + 1 - k, -\alpha - k + \tfrac{1}{2} \end{matrix} ; 1 \right)$$

$$= \frac{\Gamma(n + 1 - k)\Gamma(k + \alpha + \tfrac{1}{2})}{\Gamma(n + 1)\Gamma(\alpha + \tfrac{1}{2})}$$

$$\cdot {}_3F_2 \left(\begin{matrix} -k, (m - k - n - \alpha)/2, (m - k - n - \alpha + 1)/2 \\ -k - \alpha + \tfrac{1}{2}, m + 1 - k \end{matrix} ; 1 \right).$$

The coefficient multiplying this $_3F_2$ is positive and in the series for the $_3F_2$ there are terms with factors

(6.37)
$$\frac{(-k)_j}{(-k - \alpha + \tfrac{1}{2})_j}$$

and

(6.38) $((m - k - n - \alpha)/2)_j((m - k - n - \alpha + 1)/2)_j = 2^{-2j}(m - k - n - \alpha)_{2j}.$

The factor (6.37) is positive for $\alpha > -\tfrac{1}{2}, j = 0, 1, \cdots, k$ and the factor (6.38) is positive for $j = 0, 1, \cdots, (n + k - m + \alpha)/2$. When $\alpha = 0, 1, \cdots$, it vanishes for $2j > n + k - m + \alpha$, and this completes the proof of Theorem 6.3 for $\alpha = 0, 1, \cdots$. The deeper result that it holds for $\alpha \geq (-5 + \sqrt{17})/2$ was proved by finding a recurrence relation in n for the hypergeometric function on the right-hand side of (6.36). A method developed by Bailey [4] to extend results of Watson [3] was used. The details are fairly complicated, and so will not be given here (see Askey–Gasper [4]).

A few further results have been obtained. Theorem 6.1 fails for $\alpha < -\tfrac{1}{2}$ in a very strong sense, for if $-1 < \alpha < -\tfrac{1}{2}$ there is no $\rho < \infty$ for which

(6.39)
$$\int_0^\infty L_k^\alpha(x)L_m^\alpha(x)L_n^\alpha(x)x^\alpha e^{-\rho x} \, dx \geq 0, \qquad k, m, n = 0, 1, \cdots.$$

Also (6.39) fails for $\rho < \frac{3}{2}$ for any α and some k, m, n. Both of these results are in Askey–Gasper [4].

When $\rho = \frac{3}{2}$ there is a weaker result which follows from a result to be mentioned in Lecture 7 (see Askey [12]).

THEOREM 6.4. *If* $\alpha = -\frac{1}{2}, 0, \frac{1}{2}, \cdots$, *then*

$$\frac{1}{\{(1-r)(1-s)(1-t)[(1-r)(1-s)(1+t) \atop + (1-r)(1+s)(1-t) + (1+r)(1-s)(1-t)]\}^{\alpha+1}}$$

$$= \sum_{k,m,n=0}^{\infty} c^{\alpha}(k,m,n) r^k s^m t^n$$

with $c^{\alpha}(k, m, n) \geq 0$, $k, m, n = 0, 1, \cdots$.

Once again the proof is surprising, for it again reverts to a problem on Jacobi polynomials. And part of the proof was suggested by the proof of Theorem 6.1. The restriction on α should probably be $\alpha \geq -\frac{1}{2}$, and the extraneous condition that 2α is an integer arises because of our lack of knowledge about Jacobi polynomials, and ultimately about $_4F_3$'s. This will be given in Lecture 7.

It is unclear how these results will ultimately be viewed. There are probably combinatorial ways of looking at them, at least in the case of rational functions, and this should be investigated. There are interesting Banach algebras associated with Theorems 6.1 and 6.3. And finally there is the problem of finding the right geometric space on which these functions live. Peetre [1] solved this question for $e^{-x/2}L_n(x)$ as was mentioned in Lecture 4. Since $e^{-x}L_n(x)$ behaves like a set of characters in the sense that products can be written as a sum with positive coefficients, there is probably some space and associated group on which they live. It would be interesting to find it.

Since Laguerre polynomials are limits of Meixner polynomials it would be reasonable to suspect that there is a corresponding theorem for Meixner polynomials. This is totally false for if

$$(6.40) \qquad \sum_{x=0}^{\infty} a(x) M_n(x; \beta, c) M_k(x; \beta, c) \frac{(\beta)_x}{x!} c^x \geq 0, \qquad k, n = 0, 1, \cdots,$$

and

$$|a(x)| \leq K, \qquad x = 0, 1, \cdots,$$

then

$$a(x) = \int_0^{\infty} e^{-xt} \, d\mu(t), \quad d\mu(t) \geq 0, \qquad x = 0, 1, \cdots.$$

The weaker result that

$$\sum_{x=0}^{\infty} a(x) M_l(x; \beta, c) M_k(x; \beta, c) M_n(x; \beta, c) \frac{(\beta)_x}{x!} c^x \geq 0,$$

$$a, l, n = 0, 1, \cdots,$$

implies that $a(x) \equiv 0$, $x = 1, 2, \cdots$ was proved in Askey–Gasper [4], and the above analogue of Sarmanov's theorem for Meixner polynomials was proved by Tyan [1].

Part of the fascination of Szegö's result is that it only holds for Laguerre polynomials, or to be precise, no other polynomials whose spectral measure has unbounded support have been shown to have a similar type of positive integral. As we saw above, when $\alpha \geq (-5 + \sqrt{17})/2$ there is an improvement possible in Szegö's theorem, and a similar improvement probably exists for each $\alpha > -\frac{1}{2}$. It is likely that the integral (6.13) is strictly positive for $\alpha > -\frac{1}{2}$, and that the factor e^{-3x} can be replaced by $e^{-c(\alpha)x}$, $c(\alpha) < 3$ when $a > -\frac{1}{2}$. It is unclear if e^{-3x} is the best that can be used when $\alpha = -\frac{1}{2}$. It would be interesting to obtain the best possible result in this case. Also there are interesting problems related to rational functions. One which has caused me many hours of frustration is trying to prove that

$$(6.41) \qquad \frac{1}{\begin{array}{c}(1 - r)(1 - s)(1 - t)(1 - u)[(1 - r)(1 - s) + (1 - r)(1 - t) \\ + (1 - r)(1 - u) + (1 - s)(1 - t) + (1 - s)(1 - u) + (1 - t)(1 - u)]\end{array}}$$

has positive power series coefficients. So far the most powerful method of treating problems of this type is to translate them into another problem involving special functions and then use the results and methods which have been developed for the last two hundred years to solve the special function problem. So far I have been unable to make a reduction in (6.41) and so have no place to start. However, it is possible to solve some problems without using special functions, so others should not give up on (6.41).

Added in proof. I have been reading some of the early papers on hypergeometric series and need to modify some of the comments in this lecture. Saalschütz had been anticipated by Pfaff [2], so I propose to rename the series which Whipple called Saalschützian. In the future they should be called balanced and, more generally, k-balanced if the sum of the numerator parameters plus k is the sum of the denominator parameters and the series terminates. See Askey [18]. The formula (6.35) was found by Pfaff [1] and the proof of it given on page 53 was given by Jacobi [1].

LECTURE 7

Connection Coefficients

The last of the four basic questions to be considered is the problem of connection coefficients between two sequences of functions. If $\{p_n(x)\}$ and $\{q_n(x)\}$ are given $(n = 0, 1, \cdots)$ we wish to find the coefficients $c_{k,n}$ satisfying

$$(7.1) \qquad q_n(x) = \sum_{k=0}^{\infty} c_{k,n} p_k(x).$$

Usually (but not always) $p_n(x)$ and $q_n(x)$ are polynomials of degree n, in which case there is no question of the existence of $c_{k,n}$. In this degree of generality nothing useful can be said about the connection coefficients, and in all instances I know, very little of any interest can be said unless the sets of functions are similar, a notion which will not be made precise. For example, when considering orthogonal polynomials the true intervals of orthogonality should be the same.

The first instance of connection coefficients goes back to Stirling [1]. He defined two sets of numbers, Stirling numbers of the first and second kind:

$$(7.2) \qquad x(x - 1) \cdots (x - n + 1) = \sum_{k=0}^{n} S_n^{(k)} x^k$$

gives those of the first kind, and

$$(7.3) \qquad x^n = \sum_{k=0}^{n} \mathscr{S}_n^{(k)} x(x - 1) \cdots (x - k + 1),$$

those of the second kind. There is no agreement on a standard notation, so that in the standard handbook of Abramowitz and Stegun [1] is the notation used above. These numbers are useful in combinatorial problems, but they do not play a role in the problems in these lectures and so will not be mentioned further.

For the classical polynomials there are two very old results which can be derived easily from generating functions:

$$(7.4) \qquad L_n^{\alpha + \beta + 1}(x) = \sum_{k=0}^{n} \frac{(\beta + 1)_{n-k}}{(n - k)!} L_k^{\alpha}(x),$$

$$(7.5) \qquad C_n^{\lambda}(\cos \theta) = \sum_{k=0}^{n} \frac{(\lambda)_{n-k}(\lambda)_k}{(n - k)!k!} \cos (n - 2k)\theta.$$

The required generating functions are

$$(7.6) \qquad (1 - r)^{-\alpha - 1} \exp(-xr/(1 - r)) = \sum_{n=0}^{\infty} L_n^{\alpha}(x) r^n$$

and

$$(7.7) \qquad (1 - 2r \cos \theta + r^2)^{-\lambda} = \sum_{n=0}^{\infty} C_n^{\lambda}(\cos \theta) r^n.$$

An interesting generalization of generating functions of the type (7.6) has recently been given by Rota and coworkers (see Mullin–Rota [1] and Rota–Kahaner–Odlyzko [1]).

They define polynomials of binomial type as polynomials which satisfy

$$(7.8) \qquad p_n(x + y) = \sum_{k=0}^{n} \binom{n}{k} p_k(x) p_{n-k}(y).$$

All such polynomials have formal generating functions of the form

$$(7.9) \qquad \sum_{n=0}^{\infty} \frac{p_n(x) r^n}{n!} = e^{xf(r)}, \quad f(0) = 0, \quad f'(0) \neq 0.$$

Examples include $p_n(x) = x^n$ (hence the name binomial type) and the polynomials $\mathcal{L}_n^{-1}(x) = n! L_n^{-1}(x)$, which are defined by the generating function (7.6) when $\alpha = -1$.

More general polynomials are the Sheffer sets $\{s_n(x)\}$ relative to a binomial set $\{p_n(x)\}$. They satisfy

$$(7 \cdot 10) \qquad s_n(x + y) = \sum_{k=0}^{n} \binom{n}{k} s_k(x) p_{n-k}(y)$$

and have the generating function

$$(7.11) \qquad \sum_{n=0}^{\infty} \frac{s_n(x) r^n}{n!} = g(r) e^{xf(r)}, \qquad\qquad g(0) \neq 0,$$

and the same $f(r)$ as in (7.9). For two Sheffer sets, Rota–Kahaner–Odlyzko [1] found the connection coefficients by use of umbral calculus (or symbolic methods) which they justify by means of linear operators (see also Brown–Goldberg [1]). There are many interesting formulas in these papers, and these techniques should be learned by anyone who wants to work with polynomial sets defined by simple exponential type generating functions. It is quite likely that the most interesting applications will come from considering polynomials in several variables (see Parrish [1] and the sections in Rota–Kahaner–Odlyzko [1] on cross-sequences and Steffensen sequences).

Our main interest is Jacobi polynomials, and they are not included in the above theory. Formula (7.5) is an instance of the explicit construction of connection coefficients between two Jacobi polynomials. For Legendre polynomials this formula was proved by Laplace [1] and his proof extends immediately to ultraspherical polynomials once the generating function (3.16) is given. One simple application of (7.5) is the following:

$$(7.12) \qquad |C_n^{\lambda}(\cos \theta)| \leq \sum_{k=0}^{n} \frac{(\lambda)_{n-k} (\lambda)_k}{(n-k)! k!} = C_n^{\lambda}(1), \qquad\qquad \lambda > 0.$$

Gegenbauer [1] stated the more general formula

(7.13)
$$C_n^\lambda(x) = \sum_{k=0}^{[n/2]} a_{k,n} C_{n-2k}^\mu(x),$$

with

(7.14)
$$a_{k,n} = \frac{\Gamma(\mu)(n - 2k + \mu)\Gamma(k + \lambda - \mu)\Gamma(n - k + \lambda)}{\Gamma(\lambda)\Gamma(\lambda - \mu)k!\Gamma(n - k + \mu + 1)}.$$

Initially he stated this only when $\lambda - \mu$ was an integer, but later [3] he states it for general λ and μ. When the limit as μ tends to zero is taken and

(7.15)
$$\lim_{\mu \to 0} \left(\frac{n + \mu}{\mu}\right) C_n^\mu(\cos \theta) = \begin{cases} 1, & n = 0, \\ 2 \cos n\theta, & n = 1, 2, \cdots, \end{cases}$$

is used, then (7.5) is obtained. While (7.5) is easy to prove, it is a little harder to prove (7.13). There are three reasonably natural ways to prove it. The most straightforward is to write $C_n^\lambda(x)$ as a polynomial in x^k, and then expand x^k in a series of $C_j^\mu(x)$. The resulting coefficients can be evaluated by the Chu–Vandermonde sum

(7.16)
$$_2F_1(-n, a; c; 1) = \frac{(c - a)_n}{(c)_n}$$

(see Hua [1, § 7.1] for the details of this argument, but not for the reference to Chu or Vandermonde). As is customary in this field, he derived this formula directly, and he was probably unaware of the long history behind (7.16). Until recently I was also unaware exactly how old (7.16) is. The usual reference is Vandermonde, with occasionally the date 1770 attached. Knuth [1] gave a more precise reference (Vandermonde [1]), but Gasper called my attention to a comment in Chrystal [1]. Chrystal says that (7.16) is usually attributed to Vandermonde, but it is older. How much older he does not say, but I would guess that he thought Euler had it earlier. I have been unable to find it in Euler's work before about 1775, but this is not a very strong proof that he did not have it before 1772. However, Chrystal was right— it had been published by Chu [1] in 1303. I have not seen this fascinating book, but Needham [1, p. 138] gives the sums

$$\sum_{r=1}^{n} \frac{r^{|p|}}{1^{|p|}} \frac{(n + 1 - r)^{|q|}}{1^{|q|}} = \sum_{r=1}^{n} \frac{r^{|p+q|}}{1^{|p+q|}}$$

and

$$\sum_{r=1}^{n} \frac{r^{|p|}}{1^{|p|}} = \frac{n^{|p+1|}}{q^{|p+1|}},$$

where

$$r^{|p|} = (r)_p = \Gamma(p + r)/\Gamma(r).$$

In more standard notation these sums are

$$\frac{\Gamma(n+q)}{\Gamma(n)\Gamma(q+1)}\sum_{k=0}^{n-1}\frac{(-n+1)_k(p+1)_k}{(-n-q+1)_k k!} = \sum_{k=0}^{n-1}\frac{(p+q+1)_k}{k!}$$

and

$$\sum_{k=0}^{n-1}\frac{(p+1)_k}{k!} = \frac{\Gamma(n+p+1)}{\Gamma(n)\Gamma(p+2)}.$$

Putting them together gives

$$\sum_{k=0}^{n-1}\frac{(-n+1)_k(p+1)_k}{(-n-q+1)_k k!} = \frac{(p+2)_{n-1}}{(q+1)_{n-1}} = \frac{\Gamma(n+p+1)\Gamma(q+1)}{\Gamma(p+2)\Gamma(n+q)}.$$

It is likely that Chu only had this sum for integer values of p and q, but it is easy for us to conclude the same equality for complex p and q from his result. For both sides are rational functions of p and q which agree infinitely often. Thus Chu really had the value of the general polynomial $_2F_1$ when $x = 1$. He also had a special case of Saalschütz's formula (see Takács [1] and Carlitz [1]). Since most mathematical historians have missed these important results in Chu [1], this book should be translated so that mathematicians who cannot read Chinese can see what other treasures are contained in it. The fact that Chu had the "Pascal triangle" property of binomial coefficients is not surprising. It is a fairly obvious fact once the binomial coefficients are discovered. The Chu–Vandermonde sum (7.16) is much deeper, and not at all obvious. The distinction between these two results is really the difference between

(7.17) $(1 + x)^a(1 + x) = (1 + x)^{a+1}$

and

(7.18) $(1 + x)^a(1 + x)^b = (1 + x)^{a+b}.$

This seems a small difference, but to obtain (7.16) from this sum one must also know how to multiply polynomials of arbitrary degree and collect terms. This is far from obvious until adequate notation has been developed. And the special case of Saalschütz's formula that Chu had was absolutely incredible. To see this one only need look at the contortions some very good mathematicians went through to prove this in the middle of the twentieth century (see the papers in the bibliography of Takács [1]). Chu did not have the benefit of integral or differential calculus, the tools used by most of these people. He must have been a remarkable mathematician.

My favorite proof of (7.13) is first to calculate the coefficients in

(7.19) $P_n^{(\gamma,\beta)}(x) = \sum_{k=0}^{n} a_{k,n} P_k^{(\alpha,\beta)}(x)$

and then use the quadratic transformations

(3.13) $\frac{P_n^{(\alpha,-1/2)}(2x^2 - 1)}{P_n^{(\alpha,-1/2)}(1)} = \frac{P_{2n}^{(\alpha,\alpha)}(x)}{P_{2n}^{(\alpha,\alpha)}(1)} = \frac{C_{2n}^{\alpha+1/2}(x)}{C_{2n}^{\alpha+1/2}(1)}$

and

(3.14)
$$\frac{xP_n^{(\alpha,1/2)}(2x^2 - 1)}{P_n^{(\alpha,1/2)}(1)} = \frac{C_{2n+1}^{\alpha+1/2}(x)}{C_{2n+1}^{\alpha+1/2}(1)}$$

on the series (7.14) when $\beta = \pm\frac{1}{2}$ to derive (7.13). The connection coefficients in (7.19) are very easy to derive by orthogonality and Rodriques' formula (2.1). Explicitly they will be given in (7.33). This method has the disadvantage of having to break a problem into two cases when it should not be necessary to do this, but that is a minor objection.

When x is set equal to one in (7.19) the resulting formula gives a special case of Dougall's formula. It is

(7.20) $_5F_4\left(\begin{matrix} 2a, a + 1, b + a + 1, c + a + 1, -n \\ a, a - b, a - c, n + 2a + 1 \end{matrix}; 1\right) = \frac{(1 + 2a)_n(1 - b - c)_n}{(1 + a - b)_n(1 + a - c)_n}.$

This gives a partial explanation of a fact which has interested me for years. Well-poised series are series

(7.21)
$$\sum_{n=0}^{\infty} \frac{(a_1)_n(a_2)_n \cdots (a_p)_n}{(1)_n(a_1 - a_2 + 1)_n \cdots (a_1 - a_p + 1)_n} x^n$$

in which numerator and denominator factors can be paired so that their sums are constant. After Kummer's sum of the well-poised $_2F_1$ at $x = -1$ and Dixon's sum of the well-poised $_3F_2$ at $x = 1$, most of the well-poised series which can be summed are what I like to call "very well-poised", one of the numerator parameters is one more than the corresponding denominator parameter. This comes in very naturally from the orthogonality relation for Jacobi polynomials. For

(7.22)
$$\int_{-1}^{1} [P_n^{(\alpha,\beta)}(x)]^2(1 - x)^\alpha(1 + x)^\beta \, dx = \frac{2^{\alpha+\beta+1}\Gamma(n + \alpha + 1)\Gamma(n + \beta + 1)}{(2n + \alpha + \beta + 1)\Gamma(n + 1)\Gamma(n + \alpha + \beta + 1)}$$
$$= \frac{(\alpha + 1)_n(\beta + 1)_n((\alpha + \beta + 1)/2)_n 2^{\alpha+\beta+1}\Gamma(\alpha + 1)\Gamma(\beta + 1)}{(1)_n(\alpha + \beta + 1)_n((\alpha + \beta + 3)/2)_n\Gamma(\alpha + \beta + 2)}.$$

In a Jacobi series this appears in the denominator, so factors of the form $(2a)_n(a + 1)_n/(a)_n$, where $a = (\alpha + \beta + 1)/2$, tend to occur. This is only a partial explanation. These sums are fundamental results which can be approached from many ways and there is probably an explanation from each of these ways (see Burchnall–Lakin [1] for a partial explanation of this from the point of view of differential equations).

The third method generalizes the second in that one calculates the coefficients in

(7.23)
$$P_n^{(\gamma,\delta)}(x) = \sum_{k=0}^{n} a_{k,n}P_k^{(\alpha,\beta)}(x).$$

This can be done in many different ways, two of them being a use of Rodrigues' formula and the expansion first in terms of $(1 - x)^j$ and then expanding that in terms of $P_k^{(\alpha,\beta)}(x)$. The resulting coefficients are $_3F_2$'s and when $\delta = \beta$ or $\gamma = \alpha$

or $\gamma = \delta$, $\alpha = \beta$ they can be evaluated as products of gamma functions by use of standard formulas. Since this method gives a more general formula than the other two methods we shall give one version of it. The coefficients can be obtained by orthogonality,

$$(7.24) \qquad a_{k,n} = \frac{\int_{-1}^{1} P_n^{(\gamma,\delta)}(x)P_k^{(\xi,\beta)}(x)(1-x)^{\alpha}(1+x)^{\beta}\,dx}{\int_{-1}^{1} [P_k^{(\alpha,\beta)}(x)]^2(1-x)^{\alpha}(1+x)^{\beta}\,dx}.$$

Next use Rodrigues' formula

$$(7.25) \qquad (1-x)^{\alpha}(1+x)^{\beta}P_k^{(\alpha,\beta)}(x) = \frac{(-1)^k}{2^k k!}\frac{d^k}{dx^k}[(1-x)^{k+\alpha}(1+x)^{k+\beta}]$$

and the differentiation formula

$$(7.26) \qquad \frac{d}{dx}P_n^{(\alpha,\beta)}(x) = \frac{(n+\alpha+\beta+1)}{2}P_{n-1}^{(\alpha+1,\beta+1)}(x);$$

to obtain

$$(7.27) \qquad \begin{aligned} a_{k,n} &= \frac{(2k+\alpha+\beta+1)\Gamma(k+\alpha+\beta+1)(n+\gamma+\delta+1)_k}{2^{2k+\alpha+\beta+1}\Gamma(k+\alpha+1)\Gamma(k+\beta+1)} \\ &\quad \cdot \int_{-1}^{1} P_{n-k}^{(\gamma+k,\delta+k)}(x)(1-x)^{\alpha+k}(1+x)^{\beta+k}\,dx. \end{aligned}$$

There are a number of ways of evaluating the integral in (7.27). The easiest is to use the definition of Jacobi polynomials as hypergeometric functions and integrate term by term. Since there are a number of different representations as hypergeometric functions there are various $_3F_2$ representations of $a_{k,n}$. These are all derivable from any one of them by means of the Thomae–Whipple transformation formulas. These transformation formulas are very useful, as we saw in Lecture 6.

Using (2.2) in (7.27) leads to

$$(7.28) \qquad \begin{aligned} a_{k,n} &= \frac{\Gamma(k+\alpha+\beta+1)\Gamma(n+k+\gamma+\delta+1)\Gamma(n+\gamma+1)}{\Gamma(n+\gamma+\delta+1)\Gamma(k+\gamma+1)\Gamma(2k+\alpha+\beta+1)(n-k)!} \\ &\quad \cdot {}_3F_2\left(\begin{matrix} -n+k, n+k+\gamma+\delta+1, k+\alpha+1 \\ k+\gamma+1, 2k+\alpha+\beta+2 \end{matrix}; 1\right). \end{aligned}$$

When $\gamma = \alpha$ this $_3F_2$ reduces to a $_2F_1$ and so is summable by Gauss' formula

$$(7.29) \qquad {}_2F_1(a,b;c;1) = \frac{\Gamma(c)\Gamma(c-a-b)}{\Gamma(c-a)\Gamma(c-b)}, \qquad c > a+b,$$

or since it is a polynomial we only need the Chu–Vandermonde sum (7.16). When $\beta = \delta$ the $_3F_2$ is summable by Pfaff's formula

$$(7.30) \qquad {}_3F_2\left(\begin{matrix} -n, b, c \\ d, -n+b+c-d+1 \end{matrix}; 1\right) = \frac{(d-b)_n(d-c)_n}{(d)_n(d-b-c)_n}.$$

And finally when $\alpha = \beta$ and $\gamma = \delta$ the $_3F_2$ is summable by Watson's formula

$$
\begin{aligned}
(7.31) \quad & {}_3F_2\left(\begin{matrix} a,b,c \\ (a+b+1)/2, 2c \end{matrix}; 1\right) \\
&= \frac{\Gamma(\tfrac{1}{2})\Gamma((1+2c)/2)\Gamma((1+a+b)/2)\Gamma((1+2c-a-b)/2)}{\Gamma((1+a)/2)\Gamma((1+b)/2)\Gamma((1+2c-a)/2)\Gamma((1+2c-b)/2)}.
\end{aligned}
$$

Explicitly the formulas are

$$
\begin{aligned}
(7.32) \quad P_n^{(\alpha,\delta)}(x) &= \frac{(\alpha+1)_n}{(\alpha+\beta+2)_n} \\
&\quad \cdot \sum_{k=0}^{n} \frac{(-1)^{n-k}(\delta-\beta)_{n-k}(\alpha+\beta+1)_k(\alpha+\beta+2)_{2k}(n+\alpha+\delta+1)_k}{(1)_{n-k}(\alpha+1)_k(\alpha+\beta+1)_{2k}(n+\alpha+\beta+2)_k} \\
&\quad \cdot P_k^{(\alpha,\beta)}(x),
\end{aligned}
$$

$$
\begin{aligned}
(7.33) \quad P_n^{(\gamma,\beta)}(x) &= \frac{(\beta+1)_n}{(\alpha+\beta+2)_n} \\
&\quad \cdot \sum_{k=0}^{n} \frac{(\gamma-\alpha)_{n-k}(\alpha+\beta+1)_k(\alpha+\beta+2)_{2k}(n+\gamma+\beta+1)_k}{(1)_{n-k}(\beta+1)_k(\alpha+\beta+1)_{2k}(n+\alpha+\beta+2)_k} P_k^{(\alpha,\beta)}(x),
\end{aligned}
$$

$$
\begin{aligned}
(7.34) \quad P_n^{(\gamma,\gamma)}(x) &= \frac{(\gamma+1)_n}{(2\alpha+1)_n} \\
&\quad \cdot \sum_{k=0}^{[n/2]} \frac{(2\alpha+1)_{n-2k}(\gamma+1/2)_{n-k}(\alpha+3/2)_{n-2k}(\gamma-\alpha)_k}{(\alpha+1)_{n-2k}(\alpha+3/2)_{n-k}(\alpha+1/2)_{n-2k}k!} P_{n-2k}^{(\alpha,\alpha)}(x).
\end{aligned}
$$

Observe that the coefficients in (7.33) and (7.34) are nonnegative when $\gamma > \alpha > -1$, and in (7.32) they are nonnegative when $\delta = \beta - 1, \beta - 2, \cdots, \delta > -1$. The nonnegativity of these coefficients in (7.33) and (7.34) has been a useful fact (see Askey–Wainger [1], Askey–Gasper [5] and Lecture 8), so the problem of deciding when these coefficients are nonnegative for general $\alpha, \beta, \gamma, \delta$ is a natural one. Just as in the last lecture the problem reduces to the nonnegativity of a $_3F_2$.

One powerful method of attacking such problems is to set up recurrence relations. Gauss showed that the general $_2F_1(a, b; c; x)$ can be linearly connected with any two other $_2F_1$'s whose parameters differ from (a, b, c) by integers. He gave the explicit formulas for the contiguous $_2F_1$'s, i.e., those which differ in only one parameter and by one in this parameter, and the general result follows by iteration. The general $_3F_2(a, b, c; d, e; x)$ also has recurrence relations, but this time some of them connect a function and three contiguous functions. See Rainville [1] for methods of obtaining these relations. Four-term recurrence relations are not too easy to handle, so it was only after Watson [3] observed that three-term relations could be found for $_3F_2$'s when $x = 1$ that it was possible to obtain some relatively complete answers to questions about the positivity of $_3F_2$'s. We shall probably never know exactly what question Watson was trying to solve when he discovered this important fact, but a reading of his paper [3] shows that it was something connected with special Jacobi polynomials. Bailey [4] gave a more systematic treatment, and his methods were adapted in Askey–Gasper [2] to set up two three-term recurrence relations and they have been analyzed to obtain fairly complete results on the nonnegativity of the $_3F_2$'s in (7.28). Before stating this result another type of problem will be considered, since it leads to the same problem and suggests some possible answers.

Given two metric spaces X and Y we wish to see if X can be isometrically embedded in Y. This problem in metric geometry was attacked by Schoenberg in the following way. A function $f(t)$, $t \geq 0$, is said to be positive definite with respect to X if

$$\sum_{j,k=1}^{n} c_j c_k f(\rho_X(x_j, x_k)) \geq 0, \qquad n = 1, 2, \cdots,$$

for all real c_j and points $x_j \in X$. $\rho_X(x, y)$ is the distance between x and y measured by the metric on X. If X can be isometrically embedded in Y, then each function which is positive definite with respect to Y is also positive definite with respect to X. When the metric space X is the real line it is an immediate consequence of Bochner's theorem that the bounded, continuous positive definite functions are Fourier cosine transforms of positive measures of finite mass

$$f(t) = \int_0^\infty \cos xt \, d\mu(x), \quad d\mu(x) \geq 0, \qquad \int_0^\infty d\mu(t) < \infty.$$

When X is the unit circle, then

$$f(\cos \theta) = \sum_{n=0}^{\infty} a_n \cos n\theta. \ a_n \geq 0, \ \sum_{n=0}^{\infty} a_n < \infty,$$

Schoenberg [2] proved that the continuous positive definite functions on S^k, the k-dimensional unit sphere; $x_1^2 + \cdots + x_{k+1}^2 = 1$, are given by

$$(7.35) \quad f(\cos \theta) = \sum_{n=0}^{\infty} a_n C_n^\lambda(\cos \theta), \quad a_n \geq 0, \ \sum_{n=0}^{\infty} a_n C_n^\lambda(1) < \infty, \quad \lambda = (k-1)/2.$$

Since the unit circle can be isometrically embedded in S^k Schoenberg [2] remarked that

$$(7.36) \qquad C_n^\lambda(\cos \theta) = \sum_{j=0}^{n} a_j^\lambda \cos j\theta, \quad a_j^\lambda \geq 0, \quad \lambda = (k-1)/2, \ k = 2, 3, \cdots.$$

Also since S^k can be isometrically embedded in S^m when $k < m$ he also observed that

$$(7.37) \qquad C_n^\lambda(\cos \theta) = \sum_{j=0}^{n} a_j^{\lambda,\mu} C_j^\mu(\cos \theta), \quad a_j^{\lambda,\mu} \geq 0, \quad \lambda \geq \mu,$$

when 2λ and 2μ are positive integers. Observe that both of these results are clear from (7.5) and (7.13). Gegenbauer's formula (7.13) had been forgotten by the time Schoenberg wrote [2], so (7.37) was far from clear. Schoenberg's remark was one of the reasons I suspected a formula like (7.13) was true, and why I noticed it when reading Gegenbauer's paper [3]. It is not surprising that (7.13) was overlooked for so long because Gegenbauer has hundreds of formulas in his papers and very few of them are used. There may be other important formulas there which have been over-looked, but they will probably not be appreciated until they are rediscovered and used. As one further example, Feldheim's formula (3.30) for the special case $\lambda = 0$ had been given by Gegenbauer [3]. Periodic checks should be made of these old papers to see what has been rediscovered that was once known but was forgotten because no one knew how to us it. One can hope to be lucky and find a formula which is just what is needed, but my experience is that this almost never happens. You find a formula in an old paper only after you have rediscovered it.

Back to Schoenberg's problem of isometric embeddings and positive definite functions. There are other compact manifolds which are quite similar to spheres, the projective spaces over the reals, the complex numbers, the quaternions and a two-dimensional projective space over the Cayley numbers. These are all the compact two-point homogeneous spaces. They have the property that a group of motions exists which takes any pair of points (x_1, y_1) to (x_2, y_2) when dist (x_1, y_1) = dist (x_2, y_2). Bochner [1] proved that the positive definite functions on these spaces are given by

$$f(\theta) = \sum_{n=0}^{\infty} a_n \varphi_n(\cos \theta), \quad a_n \geq 0, \quad \sum_{n=0}^{\infty} a_n < \infty,$$

where $\varphi_n(\cos \theta)$ are suitably normalized zonal spherical functions. For the above manifolds these functions are Jacobi polynomials. Cartan [1] proved this for complex projective spaces and Helgason [1] gave the differential equations in general. Gangolli [1] then pointed out the following facts explicitly:

Space	Spherical function
S^d	$c_n P_n^{((d-2)/2, (d-2)/2)}(\cos \pi\theta/L)$
$P^d(R)$	$c_n P_{2n}^{((d-2)/2, (d-2)/2)}(\cos \pi\theta/(2L))$
$P^d(C)$	$c_n P_n^{((d-2)/2, 0)}(\cos \pi\theta/L)$
$P^d(H)$	$c_n P_n^{((d-2)/2, 1)}(\cos \pi\theta/L)$
$P^{16}(Cay)$	$c_n P_n^{(7, 3)}(\cos \pi\theta/L)$.

In all cases c_n is chosen to give the spherical function the normalization $\varphi_n(0) = 1$. Here L is the diameter of the space and d is the "real" dimension, that is, the dimension as a real manifold. So $P^{16}(Cay)$ has real dimension 16 and dimension 2 over the Cayley numbers. These formulas are nice except for the spherical function of $P^d(R)$, the real projective space. These are usually thought of as spheres with antipodal points identified, which is the reason for giving the spherical functions as above. However, it is clearly better to write this line as

| $P^d(R)$ | $c_n P_n^{((d-2)/2, -1/2)}(\cos \pi\theta/L)$ |

as can be done by use of the quadratic transformation (3.13).

Since the real numbers can be isometrically embedded in the complex numbers, which can be isometrically embedded in the quaternions, and then on to the Cayley numbers, it is possible to isometrically embed $P^d(R)$ in $P^{2d}(C)$, $P^{2d}(C)$ in $P^{4d}(H)$, and $P^8(H)$ in $P^{16}(Cay)$. Using Schoenberg's remark about the reverse inclusion among the positive definite functions we see that

$$(7.38) \qquad\qquad P_n^{(2\alpha+1,2\beta+1)}(x) = \sum_{k=0}^n a_{k,n} P_k^{(\alpha,\beta)}(x)$$

with $a_{k,n} \geq 0$ when $(\alpha, \beta) = (j/2, -\frac{1}{2})$, $(j + 1, 0)$ and $(3, 1)$, $j = 0, 1, \cdots$. Also since $P^d(R)$ can be isometrically embedded in $P^{d+1}(R)$, etc., this gives

$$(7.39) \qquad\qquad P_n^{(\gamma,\beta)}(x) = \sum_{k=0}^n a_{k,n} P_k^{(\alpha,\beta)}(x)$$

with $a_{k,n} \geq 0$ when $\gamma > \alpha$; and α, β, γ are suitably restricted. By (7.33) this always holds, so it was natural to conjecture that there was a generalization of (7.38). There is; it holds for $\alpha \geq \beta \geq 0$, and even stronger results are true. One of those given in Askey–Gasper [2] is as follows.

THEOREM 7.1. *Let*

$$V(\alpha, \beta) = \{(\gamma, \delta): \gamma \geq \alpha, \delta = \beta\},$$

$$W(\alpha, \beta) = \{(\gamma, \delta): \gamma + \delta \geq \alpha + \beta, \gamma - \delta \geq 2\alpha - 2\beta, \gamma\beta - \delta\alpha \geq \delta - \gamma + \alpha - \beta\},$$

and to any set U define its translation $U(j)$ by

$$U(\alpha, \beta; j) = \{(\gamma, \delta): (\gamma, \delta + j) \in U(\alpha, \beta), \delta > -1\}.$$

If $\alpha \geq |\beta|$, $\beta > -1$, then the coefficients $a_{k,n}$ in (7.19) are nonnegative when

$$(\gamma, \delta) \in \bigcup_{j=0}^{\infty} [V(\alpha, \beta; j) \cup W(\alpha, \beta; j)], \quad \delta > -1.$$

The condition $\gamma\beta - \delta\alpha \geq \delta - \gamma + \alpha - \beta$ puts (γ, δ) below the line connecting (α, β) and $(-1, -1)$. It is necessary as is easily seen by considering the case $n = 1$ in (7.17). The condition $\gamma - \delta \geq 2\alpha - 2\beta$ puts (γ, δ) below the line with slope one passing through $(2\alpha + 1, 2\beta + 1)$. When $\delta \geq \beta$ this condition is also necessary, as can be seen by using an asymptotic formula of Fields [1] and the $_3F_2$ in (7.28) (for further results see Askey–Gasper [2]).

These sums can be inverted (see Askey [1]), and the more general problem

$$(1 - x)^\sigma (1 + x)^\tau P_n^{(\gamma,\delta)}(x) = \sum_{k=0}^{\infty} a_{k,m} P_k^{(\alpha,\beta)}(x)$$

may also be interesting (see Askey–Gasper [2]).

The most interesting problem still open of this simple type is to see if $P^{4k}(H)$ can be isometrically embedded in $P^{4k}(C)$. Positive definite functions can be used to show that $P^{2k}(C)$ cannot be isometrically embedded in $P^{2k}(R)$, but they cannot be used to solve the same problem for quaterionic and complex projective spaces.

The interest in this problem comes from a desire to have a definite way of telling if one metric space can be embedded in another. When the embedding is in Hilbert space von Neumann and Schoenberg [1] have shown that the positive definite functions are a sufficiently large class of invariants to solve the isometric embedding problem. The above problem might provide a negative answer for an interesting class of metric spaces. On the other hand, if the isometric embedding of $P^{4k}(H)$ in $P^{4k}(C)$ is possible, then this says something interesting about the extension of the complex numbers to the quaternions. Either way the question has an interesting answer.

Before leaving the simple connection problem for orthogonal polynomials two results of a different type should be mentioned. Askey [5] proved a theorem giving positive connection coefficients when certain conditions on recurrence coefficients are satisfied. The proof is similar to the proof of Theorem 5.2.

Wilson [2] proved the following theorem.

THEOREM 7.2. *Let $p_n(x)$ be orthogonal with respect to $d\alpha(x)$ and $q_n(x)$ be orthogonal with respect to $d\beta(x)$. Let the highest coefficients of $p_n(x)$ and $q_n(x)$ be positive. Then*

$$q_n(x) = \sum_{k=0}^{n} a_{k,n} p_k(x)$$

with $a_{k,n} \geq 0$ if

$$\int p_n(x) p_m(x)\, d\beta(x) \leq 0, \qquad\qquad m \neq n.$$

He has applied this theorem to obtain a few results on an interesting class of discrete orthogonal polynomials which approximate Legendre polynomials better than the Hahn polynomials $Q_n(x; 0, 0, N)$ (see Wilson [3]).

In Lecture 6 we used

(7.40) $$(1 + x)P_n^{(\alpha,\beta + 1)}(x) = c_n P_{n+1}^{(\alpha,\beta)}(x) + d_n P_n^{(\alpha,\beta)}(x)$$

with $c_n > 0$, $d_n > 0$ when $\alpha, \beta > -1$ to prove Theorem 6.1. Theorem 6.3 was stronger than Theorem 6.1 so it seems not unlikely that there is a result which is stronger than (7.40). A proof of Theorem 6.3 for $\alpha \geq 0$ could be given if

(7.41) $$(1 + x)^2 P_n^{(\alpha,\beta + 3)}(x) = \sum_{k=0}^{n+2} c_{k,n} P_k^{(\alpha,\beta)}(x), \qquad\qquad c_{k,n} \geq 0,$$

could be proven for $\alpha \geq 0$. It is very unlikely that (7.41) holds, for the coefficients arise from two applications of (7.40) and one of

(7.42) $$P_n^{(\alpha,\beta + 1)}(x) = \sum_{k=0}^{n} a_{k,n} P_k^{(\alpha,\beta)}(x).$$

The coefficients in (7.42) satisfy $(-1)^{n-k} a_{k,n} \geq 0$, so there is no hope that the coefficients in (7.41) are nonnegative.

Surprisingly there is a result of this type when a slight change has been made. Recall that

(7.43)
$$P_n^{(2\alpha+1,2\beta+1)}(x) = \sum_{k=0}^{n} a_{k,n} P_k^{(\alpha,\beta)}(x),$$

with $a_{k,n} \geq 0$ when $\alpha \geq |\beta|, \beta > -1$. We can now ask if

(7.44)
$$(1+x)^j P_n^{(2\alpha+1,2\beta+2j+1)}(x) = \sum_{k=0}^{n+j} a_{k,n} P_k^{(\alpha,\beta)}(x)$$

with $a_{k,n} \geq 0$ when $\alpha \geq |\beta|, \beta > -1, j = 0, 1, \cdots$. In general this is still open, but it is true for $\beta = \pm\frac{1}{2}, \alpha = -\frac{1}{2}, 0, \frac{1}{2}, \cdots$. An equivalent result with an ultraspherical polynomial on the right was proved by Dunkl [1]. His argument is similar to Schoenberg's argument when he proved (7.37), but he used the unitary group rather than the orthogonal group to obtain the Jacobi polynomials on the left side of (7.44). Theorem 6.4 is an immediate consequence of this result (see Askey [12]). Since it would be interesting to prove Theorem 6.4 for $\alpha \geq -\frac{1}{2}$, it would be useful if (7.44) could be proved without the restrictions put on by use of group theoretic methods. The next step is clearly to translate this problem into a problem on hypergeometric functions. A straightforward calculation shows that $a_{k,n}$ is a positive multiple of

(7.45)
$$(-1)^n {}_4F_3\left(\begin{matrix} -n, n+2j+2\alpha+2\beta+3, j+1, j+\beta+1 \\ 2j+2\beta+2, j+k+\alpha+\beta+2, j-k+1 \end{matrix}; 1\right), \qquad k \leq j,$$

and

(7.46)
$$(-1)^N {}_4F_3\left(\begin{matrix} -N, N+2k+2\alpha+2\beta+3, k+\beta+1, k+1 \\ 2k+\alpha+\beta+2, j+k+2\beta+2, k-j+1 \end{matrix}; 1\right),$$
$$j \leq k, \quad N = n+j-k \geq 0.$$

From Dunkl's result we know these ${}_4F_3$ have the right sign behavior when $\beta = \pm\frac{1}{2}$, $\alpha = -\frac{1}{2}, 0, \frac{1}{2}, \cdots$. For the application to Theorem 6.4 it is sufficient to choose any value of β. The choice $\alpha = \beta$ is a natural one, for then (7.43) and (7.44) are identical. When $\alpha = -1$ the ${}_4F_3$'s become Saalschützian, and Whipple's transformation of a Saalschützian ${}_4F_3$ to a well-poised ${}_7F_6$ can be used to rewrite the sum in many different ways. But I have no faith in the more general conjecture that (7.44) holds when $\alpha = -1$. However, this seems to be a very reasonable conjecture when $\alpha \geq -\frac{1}{2}$, and it should be considered in some detail. At present we know very little about a ${}_4F_3$ which is not well-poised or Saalschützian, and this problem might be a wedge which helps us ask the right questions about a more general ${}_4F_3$.

The first ${}_3F_2$ which was proven to be nonnegative without either evaluating it or it being a ${}_3F_2$ with only positive terms was

(7.47)
$${}_3F_2\left(\begin{matrix} -m, -n, -\alpha-1 \\ -m-\alpha, -n-\alpha \end{matrix}; 1\right) > 0,$$
$$\alpha > 0, \quad m, n = 0, 1, \cdots.$$

This was treated by Lorentz and Zeller [1]. There may now be enough different special cases so that there is some hope of synthesizing them. However, it is unlikely that the general $_3F_2$ will ever be treated since there are too many parameters. Previous work on the location of zeros of $_pF_q$'s is summarized in Szegö [9, Chap. 6] and Hille [1]. The location of zeros of $_0F_q$, $_1F_1$ and $_2F_1$ is fairly well understood. Only a start has been made on the general $_1F_2(a, b, c; x)$ but this looks possible (see Steinig [1], Gasper [8], Makai [2], Askey–Steinig [1]). Beyond this there is little hope of solving the general problem, but it looks as if this type of problem will continue to occur, so new methods will have to be developed.

Added in proof. Feldheim [2] contains a number of formulas of the type given in this lecture and a few of a more general type. For example, he gives the connection formula between $P_n^{(\gamma, \delta)}(1 - \mu(1 - x))$ and $P_k^{(\alpha, \beta)}(x)$. A new proof and an extension of the Lorentz-Zeller inequality (7.47) is given in Askey, Gasper and Ismail [1].

LECTURE 8

Positive Sums

Finally we return to the type of sums considered in the first lecture. With very few exceptions all the sums considered below are strictly positive for $-1 < x < 1$ rather than just nonnegative, but to simplify the presentation we shall usually only state the weaker form of the inequality. I know very few applications of the stronger inequalities which do not follow from the weaker inequalities so it does not seem worthwhile to complicate the statements. By and large the proofs give the stronger inequalities.

First we shall prove the inequality of Kogbetliantz [1] which was mentioned in Lecture 1.

He proved that the $(C, 2v + 1)$ means of

$$
\text{(8.1)} \qquad \sum_{n=0}^{\infty} \frac{(n + v)}{v} \frac{C_n^v(x)C_n^v(y)}{C_n^v(1)}
$$

are nonnegative for $-1 \leq x, y \leq 1, v > 0$.

Recall that the (C, γ) means of $\sum_{n=0}^{\infty} a_n$ are defined by

$$
\text{(8.2)} \qquad \sigma_n^\gamma = \frac{n!}{(\gamma + 1)_n} \sum_{k=0}^{n} \frac{(\gamma + 1)_{n-k}}{(n - k)!} a_k.
$$

Using Gegenbauer's product formula (4.10) it is sufficient to prove

$$
\text{(8.3)} \qquad \sum_{k=0}^{n} \frac{(2v + 2)_{n-k}}{(n - k)!} \frac{(k + v)}{v} C_k^v(x) \geq 0, \qquad -1 \leq x \leq 1, \quad v > 0.
$$

Multiply (8.3) by r^n and sum, using

$$
\sum_{n=0}^{\infty} \frac{r^n(n + v)}{v} C_n^v(x) = \frac{1 - r^2}{(1 - 2xr + r^2)^{v+1}},
$$

to obtain

$$
\text{(8.4)} \qquad
\begin{aligned}
&\sum_{n=0}^{\infty} r^n \sum_{k=0}^{n} \frac{(2v + 2)_{n-k}}{(n - k)!} \frac{(k + v)}{v} C_k^v(x) \\
&= \frac{1 - r^2}{(1 - r)^{2v+2}(1 - 2xr + r^2)^{v+1}} \\
&= \frac{1 - r^2}{(1 - r)^2(1 - 2xr + r^2)} \cdot \frac{1}{(1 - r)^{2v}(1 - 2xr + r^2)^v}.
\end{aligned}
$$

71

A function is called absolutely monotonic if its power series has nonnegative coefficients. Since the product of absolutely monotonic functions is absolutely monotonic it is sufficient to show that

$$f(r) = \frac{1 - r^2}{(1 - r)^2(1 - 2rx + r^2)}$$

and

$$[g(r)]^\nu = \frac{1}{(1 - r)^{2\nu}(1 - 2xr + r^2)^\nu}, \qquad \nu > 0,$$

are absolutely monotonic for $-1 \leq x \leq 1$. The absolute monotonicity of $f(r)$ is Fejér's inequality (1.1), while the absolute monotonicity of $g_\nu(r)$ follows from the simple argument given next.

Let

$$h(r) = \log g(r).$$

Then

$$h'(r) = \frac{2}{1 - r} + \frac{2x - 2r}{1 - 2xr + r^2} = \sum_{n=0}^{\infty} [2 + 2\cos(n + 1)\theta]r^n, \qquad (x = \cos\theta),$$

so

$$h(r) = \sum_{n=1}^{\infty} \frac{[2 + 2\cos n\theta]r^n}{n}.$$

Thus $h(r)$ and finally $[g(r)]^\nu = \sum_{n=0}^{\infty} \nu^n [h(r)]^n/n!$ are absolutely monotonic.

Kogbetliantz's original proof of (8.3) is a horror (as first proofs often are), and I cannot follow it. I suspect that the claims he makes in his proof are true, but I do not know how to justify them. The proof given above is from Askey–Pollard [1]. The idea of using infinite divisibility is fairly old, going back to P. Levy, Bochner and Schoenberg [1]. P. Bateman called a paper of Rankin [1] to my attention in which Rankin obtains an analogue for power series of a theorem of Schoenberg for Laplace transforms. For power series this theorem says that $[f(r)]^\nu$ is absolutely monotonic for all $\nu > 0$ if and only if $\log f(r)$ is absolutely monotonic. So the simple device we used on $[g(r)]^\nu$ only worked because all of the positive powers were absolutely monotonic. The idea of considering $f(r)$ and $[g(r)]^\nu$ separately is due to Fejér [3]. However, he knew a simple way of proving the absolute monotonicity of $[g(r)]^{1/2}$ (his proof was given in Lecture 3) and he was not interested in fractional values of ν. Later, Fejér [4] became interested in ultraspherical sums and proved that

(8.5) $$\sum_{k=0}^{n} C_k^\nu(x) \geq 0, \qquad -1 \leq x \leq 1, \quad 0 < \nu \leq \tfrac{1}{2}.$$

The absolute monotonicity of $[g(r)]^v$ is equivalent to

(8.6)
$$\sum_{k=0}^{n} \frac{(2v)_{n-k}}{(n-k)!} C_k^v(x) \geq 0, \qquad -1 \leq x \leq 1, \quad v > 0,$$

which is stronger than (8.5) for $0 < v < \frac{1}{2}$.

Both (8.3) and (8.6) are best possible inequalities in the sense that any lower order Cesàro mean would be negative for $x = -1$ when n is odd.

Following Askey–Gasper [5] apply the results from Lecture 3 to these sums. Theorem 3.2 gives

(8.7)
$$\sum_{k=0}^{n} \frac{(2v)_{n-k}(2v)_k}{(n-k)!k!} \frac{C_k^\lambda(x)}{C_k^\lambda(1)} \geq 0, \qquad -1 \leq x \leq 1, \quad \lambda \geq v > 0.$$

Recall that

$$\frac{C_n^1(\cos \theta)}{C_n^1(1)} = \frac{\sin (n+1)\theta}{(n+1)\sin \theta}$$

so (8.7) contains the inequality

(8.8)
$$\sum_{k=0}^{n} \frac{(a)_{n-k}(a)_k}{(n-k)!k!} \frac{\sin (k+1)\theta}{(k+1)\sin \theta} \geq 0, \qquad 0 < a \leq 2, \quad 0 \leq \theta \leq \pi,$$

which for $a = 2$ reduces to Lukács' inequality

(1.16)
$$\sum_{k=0}^{n} (n+1-k)\sin (k+1)\theta \geq 0, \qquad 0 \leq \theta \leq \pi,$$

and for $a = 1$ reduces to the Fejér–Jackson–Gronwall inequality

(1.6)
$$\sum_{k=0}^{n} \frac{\sin (k+1)\theta}{k+1} \geq 0, \qquad 0 \leq \theta \leq \pi.$$

This is the only proof that I know of (8.8), which is a bit surprising. There are at least twenty proofs of (1.6), so others will probably be found for (8.8). I hope so, for it is likely that (8.8) also holds for $2 < a \leq 3$. A proof for the case $a = 3$ is given in Askey [14]. This is the inequality (1.31) which was mentioned in Lecture 1.

Part of the interest in (8.8) for $a = 3$ is that it is best possible in a new sense. Consider the formal sum

(8.9)
$$\sum_{k=0}^{\infty} \frac{(3+\varepsilon)_k}{k!} \frac{\sin (k+1)\theta}{(k+1)\sin \theta}$$

and try to find the lowest order Cesàro mean which will be nonnegative. It does not exist. No Cesàro mean is nonnegative for all $\theta, 0 \leq \theta \leq \pi$ and all $n, n = 1, 2, \cdots$. In fact,

(8.10)
$$\int_0^\infty e^{-nt} t^{1+\varepsilon} \sin xt \, dt$$

changes sign in x for some $\eta > 0$ when $\varepsilon > 0$. This implies that

$$(8.11) \qquad \int_0^y (y - t)^\gamma t^{1 + \varepsilon} \sin xt \, dt$$

changes sign in x for each γ when $\varepsilon > 0$. Since the Cesàro means of (8.9) can be used to set up Riemann sum approximations to (8.11) this shows that no Cesàro mean of (8.9) can be nonnegative in $0 \leq \theta \leq \theta_0$, $\theta_0 > 0$ for all n (see Askey [16] for more details).

Inequality (8.8) with $a = 3$ can be used to prove

$$(8.12) \qquad \sum_{k=0}^n \frac{(a)_{n-k}(a)_k}{(n-k)!k!} \frac{C_k^\lambda(x)}{C_k^\lambda(1)} \geq 0, \qquad\qquad 3 \leq a \leq 2\lambda + 1.$$

Next apply part (i) of Theorem 3.4 to obtain

$$(8.13) \qquad \sum_{k=0}^n \frac{(\lambda + 1)_{n-k}(\lambda + 1)_k}{(n-k)!k!} \frac{P_k^{(\alpha,\beta)}(x)}{P_k^{(\beta,\alpha)}(1)} \geq 0,$$

$$-1 \leq x \leq 1, \quad -1 < \lambda \leq \alpha + \beta, \quad \alpha \leq \beta, \quad \text{or} \quad 2 \leq \lambda \leq \alpha + \beta + 1, \quad \alpha \leq \beta.$$

When $\lambda = 0$ this recaptures much of (3.43). Since (3.43) holds for $|\beta| \leq \alpha \leq \beta + 1$ this suggests that (8.13) also holds for some values of (α, β) with $\alpha > \beta$. There are technical reasons why (8.13) cannot hold for $\lambda < 0$ when $\alpha = \frac{1}{2}$, $\beta = -\frac{1}{2}$. For

$$(8.14) \qquad \frac{(\lambda + 1)_2}{2} \sin \frac{\theta}{2} + (\lambda + 1)^2 \sin \frac{3\theta}{2} + \frac{(\lambda + 1)_2}{2} \sin \frac{5\theta}{2} < 0$$

when $\theta = 2\pi/3$ and $-1 < \lambda < 0$. However, it is reasonable to conjecture that (8.13) holds for $0 \leq \lambda \leq \alpha + \beta$ when $\alpha \geq \beta \geq -\frac{1}{2}$. Part (i) of Theorem 3.4 shows that (8.13) holds for $(\alpha - \mu, \beta + \mu)$, $\mu > 0$, if it holds for (α, β). So the most interesting values of β are small values, and in particular $\beta = -\frac{1}{2}$. When $\alpha = \frac{3}{2}$, $\beta = -\frac{1}{2}$ there are two values of λ which can be handled. When $\lambda = 1$, $\alpha = \frac{3}{2}$, $\beta = -\frac{1}{2}$, then (8.13) is equivalent to

$$(8.15) \qquad \sum_{k=0}^n \frac{(3)_{n-k}}{(n-k)!}(k + \tfrac{1}{2}) \sin (k + \tfrac{1}{2})\theta \geq 0, \qquad\qquad 0 \leq \theta \leq \pi,$$

(a summation by parts is necessary) and this inequality is Fejér's [7] inequality (1.21). When $\lambda = 0$, $\alpha = \frac{3}{2}$, $\beta = -\frac{1}{2}$, (8.13) is equivalent to

$$(8.16) \qquad \frac{d}{d\theta} \sum_{k=0}^n \frac{\sin (k + 1)\theta}{(k + 1) \sin \frac{1}{2}\theta} \leq 0, \qquad\qquad 0 \leq \theta \leq \pi.$$

This equivalence is an immediate consequence of Bateman's integral (3.7) when $\alpha = \frac{3}{2}$, $\beta = -\frac{1}{2}$, $\mu = 1$. Askey and Steinig [2] gave a direct proof of (8.16).

The next easiest cases should be $\alpha = \frac{5}{2}$, $\beta = -\frac{1}{2}$, $\lambda = 0, 1, 2$. The only one of these which has been settled is $\lambda = 0$. This was proved in Askey–Gasper [5]. Again it is equivalent to an interesting inequality about trigonometric functions,

$$(8.17) \qquad \frac{\sin (n-1)\theta}{(n-1)\sin \theta} - \frac{\sin (n+1)\theta}{(n+1)\sin \theta} \leq \frac{(3+\cos \theta)n}{n^2 - 1}\left[1 - \frac{\sin n\theta}{n \sin \theta}\right], \qquad 0 \leq \theta \leq \pi.$$

Since $|\sin n\theta / n \sin \theta| < 1$, $0 < \theta < \pi$, this inequality is stronger than Robertson's inequality (1.29).

When $\alpha = \beta + 1$ it is possible to use Theorem 3.4, part (v) to obtain new inequalities as was done in proving (3.43) and (8.7). If Conjecture 3.1 could be proved for $\alpha = \beta + 2$ and $\alpha = \beta + 3$ (the easiest cases which have not been solved) then (8.16) would imply

$$(8.18) \qquad \sum_{k=0}^{n} \frac{P_k^{(\beta + 2, \beta)}(x)}{P_k^{(\beta, \beta + 2)}(x)} \geq 0, \qquad -1 \leq x \leq 1, \quad \beta \geq -\frac{1}{2},$$

and (8.17) would imply

$$(8.19) \qquad \sum_{k=0}^{n} \frac{P_k^{(\beta + 3, \beta)}(x)}{P_k^{(\beta, \beta + 3)}(1)} \geq 0, \qquad -1 \leq x \leq 1, \quad \beta \geq -\frac{1}{2}.$$

Surprisingly the proof of (8.17) can be combined with Gegenbauer's connection relation (7.13) to give a proof of (8.19). Formula (8.18) is still open for $-\frac{1}{2} < \beta < 0$, but Theorem 3.4, part (ii) can be used to prove

$$(8.20) \qquad \sum_{k=0}^{n} \frac{P_k^{(3/2, \beta)}(x)}{P_k^{(\beta, 3/2)}(1)} \geq 0, \qquad -1 \leq x \leq 1, \quad \beta > -\frac{1}{2}.$$

These results are a complete statement of the state of our knowledge about (8.13) when $\beta < 0$ and the inequality holds. It fails for $\lambda = 0$, $\alpha < \frac{1}{2}$, $\beta = -\frac{1}{2}$ as a simple calculation shows when $n = 2$.

The next easiest case to consider is $\beta = 0$, and this is much easier. It should be, because the larger β is the weaker the inequality. In fact when $\beta = \frac{1}{2}$ inequalities like (8.11) should hold not only because the function is nonnegative, but because a simple multiple of it is monotone decreasing to zero in one direction. For example, (8.16) shows that

$$\left(\frac{1+x}{2}\right)^{1/2} \sum_{k=0}^{n} \frac{P_k^{(1/2, 1/2)}(x)}{P_k^{(1/2, 1/2)}(1)}$$

is monotone and it vanishes when $x = -1$. But knowing a result should be easier to prove and actually proving it are sometimes two completely different things. One method which works is to treat everything in sight as hypergeometric functions.

This works when $\beta = \lambda = 0$. For

$$
\begin{aligned}
\sum_{k=0}^{n} P_k^{(\alpha,0)}(x) &= \sum_{k=0}^{n} \frac{(\alpha+1)_k}{k!} \sum_{j=0}^{k} \frac{(-k)_j(k+\alpha+1)_j}{(\alpha+1)_j j!} \left(\frac{1-x}{2}\right)^j \\
&= \sum_{j=0}^{n} \frac{(\alpha+1)_{2j}}{j!(\alpha+1)_j} \left(\frac{x-1}{2}\right)^j \sum_{k=0}^{n-j} \frac{(2j+\alpha+1)_k}{k!} \\
&= \sum_{j=0}^{n} \frac{((\alpha+1)/2)_j((\alpha+2)/2)_j(2j+\alpha+2)_{n-j}}{j!(n-j)!(\alpha+1)_j} [2(x-1)]^j \\
&= \frac{(\alpha+2)_n}{n!} {}_3F_2 \left(\begin{matrix} -n, n+\alpha+2, (\alpha+1)/2 \\ \alpha+1, (\alpha+3)/2 \end{matrix}; \frac{1-x}{2}\right).
\end{aligned}
$$

(8.21)

The formulas which were used are

$$(2a)_{2j} = 2^{2j}(a)_j(a+\tfrac{1}{2})_j$$

and

(8.22)
$$\sum_{k=0}^{n} \frac{(a)_k}{k!} = \frac{(a+1)_n}{n!}.$$

The ${}_3F_2$ on the right-hand side of (8.21) has the form

(8.23)
$$
{}_3F_2 \left(\begin{matrix} 2a, 2b+1, a+b \\ a+b+1, 2a+2b \end{matrix}; \frac{1-x}{2}\right).
$$

There is an old formula of Clausen [1] (see also Bailey [2, p. 86])

(8.24)
$$
{}_3F_2 \left(\begin{matrix} 2a, 2b, a+b \\ a+b+\tfrac{1}{2}, 2a+2b \end{matrix}; t\right) = \left[{}_2F_1 \left(\begin{matrix} a, b \\ a+b+\tfrac{1}{2} \end{matrix}; t\right)\right]^2,
$$

which can be used to prove that a ${}_3F_2$ is positive. So it is reasonable to try to represent the ${}_3F_2$ in (8.23) in terms of the ${}_3F_2$ in (8.24) as either an integral with a positive kernel or a sum with positive coefficients. Since the parameters $a+b$ and $2a+2b$ are the same in both (8.23) and (8.24) this suggests trying to write $${}_2F_1 \left(\begin{matrix} 2a, 2b+1 \\ a+b+1 \end{matrix}; t\right)$$ in terms of ${}_2F_1 \left(\begin{matrix} 2a, 2b \\ a+b+\tfrac{1}{2} \end{matrix}; t\right)$ with a positive kernel and then generate the ${}_3F_2$ from the corresponding ${}_2F_1$. There are at least two ways to accomplish the first half of this task; use Gegenbauer's formula (7.13) giving the connection coefficients between two ultraspherical polynomials (the ${}_3F_2$ in (8.21) is a polynomial), or the Feldheim–Vilenkin integral (3.29). We shall use the first of these methods, since the second part of the above outline has been discussed in Lecture 3.

Two formulas which are needed are

(8.25) $${}_3F_2(a, b, c; d, e; t) = \frac{\Gamma(e)t^{1-e}}{\Gamma(c)\Gamma(e-c)} \int_0^t (t-y)^{e-c-1} y^{c-1} {}_2F_1(a, b; d; y)\, dy,$$

$$e > c > 0,$$

and

$$(7.13) \qquad C_n^\mu(x) = \sum_{j=0}^{[n/2]} \frac{(\mu - \lambda)_j (\mu)_{n-j} (\lambda + 1)_{n-2j}}{j!(\lambda + 1)_{n-j} (\lambda)_{n-2j}} C_{n-2j}^\lambda(x)$$

with $\mu = (\alpha + 2)/2$, $\lambda = (\alpha + 1)/2$. Also two representations for $C_n^\lambda(x)$ as a hypergeometric function will be used,

$$(8.26) \qquad C_n^\lambda(x) = \frac{(2\lambda)_n}{n!} {}_2F_1\left(-n, n + 2\lambda; \lambda + \tfrac{1}{2}; \frac{1-x}{2}\right),$$

$$(8.27) \qquad C_n^\lambda(x) = \frac{(2\lambda)_n}{n!} {}_2F_1\left(\frac{-n}{2}, \frac{n+2\lambda}{2}; \lambda + \tfrac{1}{2}; 1 - x^2\right).$$

The first is the standard representation and the second follows from the first by one of Kummer's quadratic transformations. Putting all these formulas together gives (when $\alpha > -1$)

$$(8.28)$$

$$\sum_{k=0}^n P_k^{(\alpha,0)}(x) = \frac{\Gamma(\alpha + 1)t^{-\alpha}}{[\Gamma((\alpha + 1)/2)]^2} \int_0^t [y(t - y)]^{(\alpha-1)/2} C_n^{(\alpha+2)/2}(1 - 2y)\, dy$$

$$= \frac{\Gamma(\alpha + 1)t^{-\alpha}}{[\Gamma((\alpha + 1)/2)]^2} \sum_{j=0}^{[n/2]} \frac{(1/2)_j ((\alpha + 2)/2)_j ((\alpha + 3)/2)_{n-2j}}{j!((\alpha + 3)/2)_j ((\alpha + 1)/2)_{n-2j}}$$

$$\cdot \int_0^t [y(t - y)]^{(\alpha-1)/2} C_{n-2j}^{(\alpha+1)/2}(1 - 2y)\, dy$$

$$= \sum_{j=0}^{[n/2]} \frac{(1/2)_j ((\alpha + 2)/2)_{n-j} ((\alpha + 3)/2)_{n-2j} (\alpha + 1)_{n-2j}}{j!((\alpha + 3)/2)_{n-j} ((\alpha + 1)/2)_{n-2j} (n - 2j)!}$$

$$\cdot {}_3F_2\left(\begin{matrix} -n + 2j, n - 2j + \alpha + 1, (\alpha + 1)/2 \\ \alpha + 1, (\alpha + 2)/2 \end{matrix}; t\right)$$

$$= \sum_{j=0}^{[n/2]} \frac{(1/2)_j ((\alpha + 2)/2)_{n-j} ((\alpha + 3)/2)_{n-2j} (n - 2j)!}{j!((\alpha + 3)/2)_{n-j} ((\alpha + 1)/2)_{n-2j} (\alpha + 1)_{n-2j}}$$

$$\cdot \left(C_{n-2j}^{(\alpha+1)/2}\left(\left(\frac{1+x}{2}\right)^{1/2}\right) \right)^2.$$

Now the condition $\alpha > -1$ can be relaxed to $\operatorname{Re}\alpha > -2$ by analytic continuation, and the coefficients of the last sum in (8.28) are positive for $\alpha > -2$ ($\alpha \neq -1$). The case $\alpha = -1$ is best handled directly, since

$$(8.29) \qquad \sum_{k=0}^n P_k^{(-1,0)}(x) = [1 + P_n(x)]/2, \quad P_n(x) = P_n^{(0,0)}(x),$$

and

$$|P_n(x)| < 1, \qquad\qquad\qquad -1 < x < 1,$$

gives

$$(8.30) \qquad \sum_{k=0}^{n} P_k^{(-1,0)}(x) > 0, \qquad\qquad -1 < x \leqq 1.$$

The above argument (due to Gasper and found in Askey–Gasper [5]) shows in a nice way how old results on hypergeometric functions can be used to prove results which are inaccessible at present by any other method. Clausen's formula was published in 1828 and Gegenbauer's formula (7.13) was published in 1884. The only other formulas used were Euler's expression of a beta function as a product of gamma functions and the sum (8.22), which is a formula of Chu [1] (see the discussion in Lecture 7).

A summary of the results which have been proved on (8.13) is contained in the following theorem.

THEOREM 8.1. *The inequality*

$$(8.13) \qquad \sum_{k=0}^{n} \frac{(\lambda + 1)_{n-k}(\lambda + 1)_k}{(n-k)!k!} \frac{P_k^{(\alpha,\beta)}(x)}{P_k^{(\beta,\alpha)}(1)} \geqq 0, \qquad -1 \leqq x \leqq 1, \quad n = 0, 1, \cdots,$$

holds when

    (I) $-1 < \lambda \leqq \alpha + \beta + 2; -\beta - 1 \leqq \alpha \leqq -2,$

    (II) $0 \leqq \lambda \leqq \alpha + \beta + 2; \alpha \leqq -2, -2 - \alpha \leqq \beta \leqq -1 - \alpha,$

    (III) $0 \leqq \lambda \leqq \beta; -2 \leqq \alpha \leqq -\beta - 1, \beta \geqq 0,$ *or* $0 \leqq \beta \leqq \alpha - 1,$

    (IV) $-1 < \lambda \leqq \beta; -2 \leqq \alpha \leqq -1, \beta \geqq -\alpha - 1,$ *or* $-1 \leqq \alpha \leqq -\frac{1}{2},$
        $-1 - \alpha \leqq \beta \leqq 1 - \alpha,$ *or* $-\frac{1}{2} \leqq \alpha \leqq \frac{1}{2}, \alpha \leqq \beta \leqq 1 - \alpha,$

    (V) $-1 < \lambda \leqq \beta$ *or* $2 \leqq \lambda \leqq \alpha + \beta + 1; -1 \leqq \alpha \leqq 0, 1 - \alpha \leqq \beta \leqq 2,$

    (VI) $-1 < \lambda \leqq \alpha + \beta + 1; -1 \leqq \alpha \leqq 0, \beta \geqq 2,$ *or* $0 \leqq \alpha \leqq 1, \beta \geqq 2 - \alpha,$
        or $1 \leqq \alpha \leqq \beta,$

    (VII) $-1 < \lambda \leqq \alpha + \beta,$ *or* $2 \leqq \lambda \leqq \alpha + \beta + 1; 0 \leqq \alpha \leqq \frac{1}{2}, 1 - \alpha \leqq \beta \leqq 2 - \alpha,$
        or $\frac{1}{2} \leqq \alpha \leqq 1, \alpha \leqq \beta \leqq 2 - \alpha,$

    (VIII) $0 \leqq \lambda \leqq \alpha + \beta; 0 \leqq \alpha \leqq \frac{1}{2}, -\alpha \leqq \beta \leqq \alpha$ *or* $\frac{1}{2} \leqq \alpha, \alpha - 1 \leqq \beta \leqq \alpha,$

    (IX) $\lambda = 0; \beta \leqq 0, 1 - \beta \leqq \alpha \leqq \frac{3}{2}$ *or* $\beta \leqq 0, 2 - \beta \leqq \alpha \leqq \beta + 3.$

One result has been omitted. Relation (8.11) also holds when $\lambda = \alpha + \beta$ and $1 \leqq \alpha \leqq \frac{3}{2}, 1 - \alpha \leqq \beta \leqq \alpha - 1,$ or $\frac{3}{2} \leqq \alpha \leqq 3, \alpha - 2 \leqq \beta \leqq \alpha - 1,$ or $3 \leqq \alpha \leqq \frac{7}{2},$ $4 - \alpha \leqq \beta \leqq \alpha - 1,$ or $\alpha \geqq \frac{7}{2}, \alpha - 3 \leqq \beta \leqq \alpha - 1.$ The statement of Theorem 8.1 is already so complicated that I did not feel like breaking up the plane into further regions. In any case it is clear that Theorem 8.1 is an interim result which is subject to change as new results are obtained. At the current rate it is likely to be out of date by the time these lectures appear in print, so it is not too important to have it stated in the most elegant way. To get some idea of what has been proved so far draw a graph of the regions in Theorem 8.1.

The one result above which is complete is (VI). When $\alpha > -1$ the inequality (8.13) does not hold when $\lambda > \alpha + \beta + 1$ (see Askey [16]).

It would be nice if the above results had been found in the order they were presented. Unfortunately that is not the case. The first problem which was discovered was the following conjecture (see Askey [10]).

CONJECTURE 8.1. *The* $(C, \alpha + \beta + 2)$ *means of*

$$\sum_{n=0}^{\infty} \frac{(2n + \alpha + \beta + 1)\Gamma(n + \alpha + \beta + 1)\Gamma(n + 1)}{\Gamma(n + \alpha + 1)\Gamma(n + \beta + 1)} P_n^{(\alpha,\beta)}(x)P_n^{(\alpha,\beta)}(1)$$

are nonnegative for $-1 \leq x \leq 1$, $\alpha \geq \beta \geq -\frac{1}{2}$. *This is equivalent to*

$$(8.31) \qquad \sum_{k=0}^{n} \frac{(\alpha + \beta + 3)_{n-k}(2k + \alpha + \beta + 1)(\alpha + \beta + 1)_k}{(n-k)!k!} \frac{P_k^{(\alpha,\beta)}(x)}{P_k^{(\beta,\alpha)}(1)} \geq 0.$$

Kogbetliantz [1] proved this when $\alpha = \beta \geq -\frac{1}{2}$. I had not understood why the $(C, 2\alpha + 2)$ means were needed to get a positive summability method when $\alpha = \beta$ when the $(C, \alpha + \frac{1}{2} + \varepsilon)$ means sum the Jacobi series of a continuous function uniformly to the function. Why a growth rate of 2α in one problem and α in the other? Finally I realized that Fejér's result (1.21) was the special case $\alpha = \frac{1}{2}, \beta = -\frac{1}{2}$ of Conjecture 8.1, and this suggested the general conjecture. The next problem was

$$(8.32) \qquad \sum_{k=0}^{n} \frac{P_k^{(\alpha,\beta)}(x)}{P_k^{(\beta,\alpha)}(1)} \geq 0, \qquad\qquad -1 \leq x \leq 1.$$

Partial results were obtained on both of these problems, so it became natural to look for intermediate problems. The clue to finding these intermediate problems was the observation that (8.31) could be summed by parts to obtain

$$(8.33) \qquad \begin{aligned} &\sum_{k=0}^{n} \frac{(\alpha + \beta + 3)_{n-k}}{(n-k)!} \frac{(2k + \alpha + \beta + 1)(\alpha + \beta + 1)_k}{k!} \frac{P_k^{(\alpha,\beta)}(x)}{P_k^{(\beta,\alpha)}(1)} \\ &\qquad = (\alpha + \beta + 1) \sum_{k=0}^{n} \frac{(\alpha + \beta + 2)_{n-k}(\alpha + \beta + 2)_k}{(n-k)!k!} \frac{P_k^{(\alpha+1,\beta)}(x)}{P_k^{(\beta,\alpha+1)}(1)}. \end{aligned}$$

Now an intermediate problem between (8.32) and (8.33) is obvious, and the proofs given above were found during the next two years.

There are limiting results and problems for Laguerre and Hermite polynomials and all of these problems can be dualized. Further details are given in Askey–Gasper [5].

Applications will be given in Lecture 9. We close this lecture with a few other problems which have come up while trying to solve these problems. As we saw in the proof of (8.3), it sometimes pays to consider generating functions and then try to prove that a certain function is absolutely monotonic. The class of absolutely monotonic functions is convex, so it is an interesting problem to find the extreme points of convex subsets. For example,

$$\frac{1}{(1 - r)^2(1 - 2xr + r^2)} = \frac{(1 + r)^2 - 2(1 + x)r}{(1 - r)^2(1 - 2xr + r^2)^2}$$

and

$$\frac{(1 + r)^2}{(1 - r)^2(1 - 2xr + r^2)^2}$$

are absolutely monotonic by results of Fejér. They are the squares of $(1 - r)^{-1}$ $\cdot (1 - 2xr + r^2)^{-1/2}$ and $(1 + r)(r - 1)^{-1}(1 - 2xr + r^2)^{-1}$, and these are the generating functions of (1.22) and (1.1). By convexity,

$$g_\lambda(r) = \frac{(1 + r)^2 + 2\lambda(1 + x)r}{(1 - r)^2(1 - 2xr + r^2)^2}$$

is absolutely monotonic for $-1 \le \lambda \le 0$. This could have suggested the problem of finding other λ for which $g_\lambda(r)$ is absolutely monotonic. It is absolutely monotonic for $-1 \le \lambda \le 1$, and not absolutely monotonic for any $\lambda > 1$ or $\lambda < -1$. As an extreme point the case $\lambda = 1$ should be interesting, and it is. It is the generating function for the function in (1.21), so its absolute monotonicity is also a result of Fejér. It took Fejér 25 years to go from $g_0(r)$ to $g_1(r)$. It is not clear if new results of interest can be found this way, but it should be considered. One test problem is to find the extreme points of the convex set

$$f(r) = \frac{A(1 + r)^4 + B(1 + r)^3 + C(1 + r)^2 + D(1 + r) + E}{(1 - r)^2(1 - 2xr + r^2)^2},$$

where $f(r)$ is absolutely monotonic and A, B, C, D, E are linear functions of x.

An interesting function arose while trying to prove (8.13) for $\alpha = \frac{3}{2}, \beta = \frac{1}{2}, \lambda = 3$. It is

$$\frac{3 + 4r - 2xr - r^2 - 4xr^2}{(1 - r)^4(1 - 2xr + r^2)^3}.$$

This is probably absolutely monotonic for $-1 \le x \le 1$, and is almost surely an extreme point in an interesting convex set of absolutely monotonic functions (see Askey [16]).

Another problem which arose when proving part (III) of Theorem 8.1 is the following: What is the smallest value c for which

$$(1 - r)^{-c\alpha}(1 - r + \sqrt{1 - 2xr + r^2})^{-\alpha}$$

is absolutely monotonic for all $\alpha > 0$ and $-1 \le x \le 1$. It is likely to be

$$c = -J_0(j_{1,1}) \cong 0.402.$$

Here $j_{1,1}$ is the first positive zero of $J_1(x)$. No $c < -J_0(j_{1,1})$ will work and this conjecture would follow from

(8.34) $$\frac{P_n(x) + P_{n-1}(x)}{2} = P_n^{(0,-1)}(x) > J_0(j_{1,1}) \cong 0.402.$$

The inequality (8.34) would follow from a conjectured set of inequalities for the relative maxima of $|P_n^{(0,-1)}(x)|$ (see Askey–Gasper [5] or Askey's comments to the solution of Problem 73.21 in SIAM Review, 16 (1974), pp. 550–552).

Finally there are useful inequalities which hold only for $0 \leq \theta \leq \theta_0$, where $0 < \theta_0 < \pi$. The first sharp one was Schweitzer's [1] inequality

$$(8.35) \qquad \sum_{k=0}^{n} \frac{(3)_{n-k}}{(n-k)!}(k+1) \sin (k+1)\theta \geq 0, \qquad 0 \leq \theta \leq \frac{2\pi}{3}.$$

Another way of stating this is that

$$(8.36) \qquad \frac{1+r}{(1-r)^2(1-2xr+r^2)^2}$$

is absolutely monotonic for $-\frac{1}{2} \leq x \leq 1$. When $0 \leq x \leq 1$ a stronger result is true:

$$(8.37) \qquad (1-r)^{-2}(1-2xr+r^2)^{-2}$$

is absolutely monotonic for $0 \leq x \leq 1$ (see Askey–Fitch [4]). It is likely that

$$(8.38) \qquad (1-r)^{-c}(1-2xr+r^2)^{-c}$$

is absolutely monotonic for $x_c \leq x \leq 1$, $x_c < 1$ when $c > 1$, and that $x_d < x_c$ when $1 < c < d$. That $x_c < 1$ is equivalent to showing that

$$(8.39) \qquad \begin{aligned} {}_3F_2(-n, n+3\gamma, \gamma; 3\gamma/2, (3\gamma+1)/2; y) &\geq 0, \\ 0 \leq y \leq y_\gamma, n &= 0, 1, \cdots, \end{aligned}$$

$y_\gamma > 0$ when $\gamma > 1$. If this is true, then it is likely that if y_γ denotes the largest value for which (8.39) holds, then $y_\gamma < y_\delta$ when $1 < \gamma < \delta$. This would follow from the following conjecture,

$$_3F_2\left(\begin{matrix} -n, n+3\delta, \delta \\ 3\delta/2, (3\delta+1)/2 \end{matrix}; y\right) = \sum_{k=0}^{n} a(k, n)\,_3F_2\left(\begin{matrix} -k, k+3\gamma, \gamma \\ 3\gamma/2, (3\gamma+1)/2 \end{matrix}; y\right)$$

with $a(k, n) \geq 0$ when $1 \leq \gamma \leq \delta$.

Added in proof. As suggested above, Theorem 8.1 was only an interim result. Gasper has proved Conjecture 8.1 when $\alpha \geq \beta + 1 \geq \frac{1}{2}$, which gives (8.13) when $\lambda = \alpha + \beta$, $\alpha \geq \beta \geq -\frac{1}{2}$. He also has proved (8.13) for $0 \leq \lambda \leq \alpha + \beta$, $\alpha \geq \beta \geq -\frac{1}{2}$. His proof uses some absolutely incredible explicit sums. In particular he has a second proof of (8.8).

LECTURE 9

More Positive Sums and Applications

One way of trying to get a better understanding of the problems in the last lecture is to see what other inequalities they imply. If these other inequalities can be proved, then this will give a further indication that the conjectures are probably true and might suggest some methods for proving them. For example, in the sum in (8.13) let $\beta = -\frac{1}{2}$, take $x = \cos \theta/n$ and use the limit relation

$$(9.1) \qquad \lim_{n \to \infty} \frac{P_n^{(\alpha,\beta)}(\cos \theta/n)}{n^\alpha} = A_\alpha \theta^{-\alpha} J_\alpha(\theta).$$

The resulting inequality is

$$(9.2) \qquad \int_0^\theta (\theta - \varphi)^\lambda \varphi^{\lambda + 1/2} J_\alpha(\varphi)\, d\varphi \geqq 0.$$

Since (8.13) probably holds for $0 \leqq \gamma \leqq \alpha + \beta, \beta \geqq -\frac{1}{2}$, it is likely that (9.2) holds for $0 \leqq \lambda \leqq \alpha - \frac{1}{2}$. The special case $\lambda = 0$ is classical, it was proved by Makai [1] by means of a very nice extension of the Sturm comparison theorem. The classical statement of Sturm's theorem concerns the oscillation of solutions of two differential equations. Watson [2, 15.83] made the very useful observation that the value of two solutions can be compared by an almost identical argument. In particular, he proved the following theorem.

THEOREM 9.1. *Let $\mu_1(x)$ and $\mu_2(x)$ be solutions of the equations*

$$\frac{d^2\mu_1}{dx^2} + I_1\mu_1 = 0, \quad \frac{d^2\mu_2}{dx^2} + I_2\mu_2 = 0$$

such that, when $x = a$,

$$\mu_1(a) = \mu_2(a), \quad \mu_1'(a) = \mu_2'(a),$$

let I_1 and I_2 be continuous in the interval $a \leqq x \leqq b$, and also let $\mu_1'(x)$ and $\mu_2'(x)$ be continuous in the same interval. Then, if $I_1 \geqq I_2$ throughout the interval, $|\mu_2(x)|$ exceeds $|\mu_1(x)|$ so long as x lies between a and the first zero of $\mu_1(x)$ in the interval, so that the first zero of $\mu_1(x)$ in the interval is on the left of the first zero of $\mu_2(x)$.

To be applicable to problems like (9.2) when $\lambda = 0$, one added refinement is necessary. The restriction that I_1 and I_2 be continuous in the interval $a \leqq x \leqq b$ needs to be replaced by continuity in $a < x \leqq b$. Szegö [5] gave this extension. Makai [1] was the first to show that this is a very powerful theorem. He not only proved (9.2) when $\lambda = 0$, he proved an old conjecture from quantum mechanics

83

that

$$\int_{x_{k,n}}^{x_{k-1,n}} |H_n(x)|^2 \, e^{-x^2} \, dx > \int_{x_{k+1,n}}^{x_{k,n}} |H_n(x)|^2 \, e^{-x^2} \, dx,$$

$$k = 1, 2, \cdots, [n/2],$$

where $x_{k,n}$ are the zeros of $H_n(x)$ ordered by $x_{k,n} > x_{k+1,n}, k = 1, 2, \cdots, n - 1$, and $x_{0,n} = \infty$.

Many other inequalities can be obtained from this theorem, but care must be taken because of the lack of continuity of the interesting differential equations at an endpoint (see Lorch [1]). Makai [2] showed how to use this type of theorem in a more sophisticated way when he proved the following inequality:

$$(9.3) \qquad\qquad \int_0^\theta \varphi^\mu J_\alpha(\varphi) \, d\varphi > 0,$$

$$-\tfrac{1}{2} < \alpha < \tfrac{1}{2}, \quad -1 - \alpha < \mu < \mu(\alpha),$$

where $\mu(\alpha)$ is defined by

$$(9.4) \qquad\qquad \int_0^{j_{\alpha,2}} \varphi^{\mu(\alpha)} J_\alpha(\varphi) \, d\varphi = 0,$$

$j_{\alpha,2}$ the second positive zero of $J_\alpha(\varphi)$. The same inequality was proved for $-1 < \alpha < -\tfrac{1}{2}$ in Askey–Steinig [1].

To see one of the basic facts behind these integrals, and the sums considered in the last lecture, consider the two integrals

$$(9.5) \qquad\qquad \int_0^x \sin t \, dt = 1 - \cos x,$$

$$(9.6) \qquad\qquad \int_0^x \cos t \, dt = \sin x.$$

The first is nonnegative, but the second is not because the first arch of $\cos t$, when $0 \le t \le \pi/2$, contains less area than the second arch of $\cos t$, when $\pi/2 \le t \le 3\pi/2$. There are two natural ways to change (9.6) so that it becomes nonnegative. One is to count the first arch more by weighting points there more than later points. One way of doing this weighting is to consider

$$(9.7) \qquad\qquad \int_0^x \frac{\cos t}{t^\gamma} \, dt \ge 0.$$

If γ is large enough, then (9.7) will hold. In particular it holds if

$$(9.8) \qquad\qquad \int_0^{3\pi/2} \frac{\cos t}{t^\gamma} \, dt = 0,$$

γ is approximately 0.3084438 (see Church [1] and Luke, Fair, Coombs and Moran [1]). This result and an inequality of Szegö

$$(9.9) \qquad \int_0^x t^{-\alpha} J_\alpha(t)\, dt > 0, \qquad\qquad x \geq 0, \quad \alpha > \bar{\alpha},$$

where

$$(9.10) \qquad \int_0^{j_{\bar{\alpha},2}} t^{-\bar{\alpha}} J_{\bar{\alpha}}(t)\, dt = 0$$

(see the appendix to Feldheim [1]), suggested the inequality (9.3).

A second way of changing (9.6) to make it nonnegative is to integrate it a second time,

$$(9.11) \qquad \int_0^x (x - t) \cos t\, dt = \int_0^x \int_0^u \cos t\, dt\, du = \int_0^x \sin t\, dt = 1 - \cos x.$$

This counts the first arch more than later arches in a different way.

For the series

$$\tfrac{1}{2} + \sum_{k=1}^n \cos k\theta$$

the second method leads to Fejér's sum (1.2), while the first method leads to a number of different sums. Rogosinski and Szegö [1] proved

$$(9.12) \qquad \tfrac{1}{2} + \sum_{k=1}^n \frac{\cos k\theta}{k+1} \geq 0, \qquad\qquad 0 \leq \theta \leq \pi.$$

Earlier, W. H. Young [1] had proved

$$(9.13) \qquad \frac{1}{1+\alpha} + \sum_{k=1}^n \frac{\cos k\theta}{k+\alpha} \geq 0, \qquad\qquad 0 \leq \theta \leq \pi,$$

for $-1 < \alpha \leq 0$, and the Rogosinski–Szegö result gives (9.13) for $0 < \alpha \leq 1$ by a summation by parts, as they remarked. Gasper [1] completed this result by finding the largest value of α for which (9.13) holds. The extreme case comes when $n = 3$ and the polynomial has a double zero. The approximate value is $\alpha = 4.5678$. At present this inequality seems to be a curiosity, and I do not know how to use it to prove any interesting results. There is another way of weighting the coefficients which also seemed to be a curiosity when it was first published. This is Vietoris' inequality

$$(9.14) \qquad \sum_{k=0}^n a_k \cos k\theta > 0, \qquad 0 \leq \theta < \pi, \quad n = 0, 1, \cdots,$$

where

$$(9.15) \qquad a_{2k} = a_{2k+1} = (\tfrac{1}{2})_k / k!, \qquad\qquad k = 0, 1, \cdots,$$

(see Vietoris [1]). It is now clear that this inequality is quite deep, and it has some interesting applications. This will be treated below after a few more inequalities have been given.

For sine series there are analogues of these inequalities. Since $\sum_{k=1}^{n} \sin k\theta$ changes sign the procedures above can be applied to suggest the inequalities

$$(9.16) \qquad \sum_{k=1}^{n} (n + 1 - k) \sin k\theta \geq 0, \qquad\qquad 0 \leq \theta \leq \pi,$$

and

$$(9.17) \qquad \sum_{k=0}^{n} \frac{\sin k\theta}{k} \geq 0, \qquad\qquad 0 \leq \theta \leq \pi.$$

These inequalities are very familiar to anyone who has read this far. Vietoris [1] found an extension of (9.17) to

$$(9.18) \qquad \sum_{k=1}^{n} a_k \sin k\theta > 0, \qquad 0 < \theta < \pi, \quad n = 1, 2, \cdots,$$

where a_k is defined by (9.15).

To see what is really behind Vietoris' inequalities (9.14) and (9.18) we return to Bessel functions and the integral (9.3). One way of thinking of this integral is as a positive multiple of the limit as $n \to \infty$ of

$$\sum_{k=0}^{n} \frac{P_k^{(\alpha,\beta)}(x)}{P_k^{(\beta,\alpha)}(1)}$$

when $x = \cos \theta/n$ and $\mu = -\beta$. But there is another way of looking at

$$\int_0^\theta \varphi^{-\mu} J_\alpha(\varphi) \, d\varphi.$$

It is also

$$\theta^{1-\mu} \int_0^1 t^{-\mu} J_\alpha(\theta t) \, dt.$$

Since

$$(9.19) \qquad x^{\mu-1} = \frac{\int_0^\infty t^{-\mu} J_\alpha(xt) \, dt}{\int_0^\infty t^{-\mu} J_\alpha(t) \, dt}, \qquad -\tfrac{1}{2} < \mu < \alpha + 1,$$

another way of looking at (9.3) is to first write $x^{\mu-1}$ as the Hankel transform (9.19) and then see when the partial integrals are positive (or nonnegative). Since the function $t^{2\alpha+1}$ is the weight function for which $t^{-\alpha} J_\alpha(t)$ are orthogonal, i.e.,

$$(9.20) \qquad \int_0^\infty \frac{J_\alpha(xt)}{(xt)^\alpha} \frac{J_\alpha(yt)}{(yt)^\alpha} t^{2\alpha+1} \, dt = \frac{\delta(x, y)}{(xy)^\alpha}$$

(see (2.45a)), this way of looking at (9.3) suggests the following problem for Jacobi polynomials.

Expand

(9.21)
$$(1 - x)^{-\gamma}(1 + x)^{-\delta} \sim \sum_{k=0}^{\infty} a_k P_k^{(\alpha,\beta)}(x)$$

and ask when

(9.22)
$$\sum_{k=0}^{n} a_k P_k^{(\alpha,\beta)}(x) \geqq 0, \qquad -1 \leqq x \leqq 1, n = 0, 1, \cdots.$$

Vietoris' inequalities (9.14) and (9.18) are the special cases

$$\alpha = \beta = -\tfrac{1}{2}, \quad \gamma = \tfrac{1}{4}, \quad \delta = -\tfrac{1}{4}$$

and

$$\alpha = \beta = \tfrac{1}{2}, \quad \gamma = \tfrac{3}{4}, \quad \delta = \tfrac{1}{4},$$

respectively. The expansions required to show this are in Vietoris [1] and this connection was pointed out in Askey–Steinig [1].

For some values of γ, δ this problem is related to an old quadrature problem. When $\gamma = \alpha$, $\delta = \beta$, then the series (9.22), when evaluated at the zeros of $P_n^{(\alpha,\beta)}(x)$, gives the Cotes' numbers for the quadrature problem of integrating the Lagrange interpolation polynomial at the zeros of $P_n^{(\alpha,\beta)}(x)$ with respect to dx (see Szegö, [9, Chap. 15]). Fejér [5], Pólya [2] and Szegö [3] studied this problem and a summary of the results up to 1938 is given in Szegö [9, Chap. 15]. Convergence holds for each continuous function if $-1 < \alpha, \beta \leqq \tfrac{3}{2}$ and there is a continuous function for which convergence fails when $\alpha > \tfrac{3}{2}$ or $\beta > \tfrac{3}{2}$. If the Cotes' numbers are positive, then convergence is easy to prove and Szegö made a start on the problem of finding out when Cotes' numbers are positive. He showed they are positive when $\alpha = \beta$, $-1 < \alpha \leqq 0$ and $\tfrac{1}{2} \leqq \alpha \leqq 1$, and they are ultimately positive for $0 < \alpha < \tfrac{1}{2}$ and $1 < \alpha \leqq \tfrac{3}{2}$. The positivity for all n in these two cases was proved in Askey–Fitch [1]. Five years ago I thought this was a fairly dull problem, to be looked at only as a test problem to see how much was known about Jacobi polynomials. After all, the convergence problem had been completely solved, and when $\alpha, \beta < \tfrac{3}{2}$ a stronger type of convergence was known (see Holló [1] and Turán [2]). But it was annoying that the only results known when $\alpha \neq \beta$ were two relatively simple cases, $\alpha = 1$, $\beta = 0$ and $\alpha = \tfrac{1}{2}, \beta = -\tfrac{1}{2}$ (see Fejér [5] and Szegö [3]). A start was made on the general case by using

$$\sum_{k=0}^{n} \frac{P_k^{(\alpha,\beta)}(x)}{P_k^{(\beta,\alpha)}(1)} \geqq 0, \qquad -\beta \leqq \alpha \leqq \beta + 1,$$

to show that (9.22) holds for $\alpha, \beta \geqq 0$, $\alpha + \beta \leqq 1$ and $\alpha = \beta + 1$, $-\tfrac{1}{2} \leqq \beta \leqq 0$. Also (9.22) fails for $\alpha > \beta + 1$ (see Askey [6]).

The general problem (9.22) for arbitrary γ, δ also has a connection with quadrature problems. The values of (9.22) at the zeros of $P_n^{(\alpha,\beta)}(x)$ are the Cotes' numbers when integrating with respect to $(1 - x)^{\alpha - \gamma}(1 + x)^{\beta - \delta} dx$. The case $\gamma = \delta = 0$ is just the classical case of Gaussian quadrature. As usual the case $\beta = -\frac{1}{2}$ is especially interesting. When $\beta = -\frac{1}{2}$, $\delta = 0$ there is convergence for each continuous function when $\gamma \leq \frac{1}{2}\alpha + \frac{3}{4}$ (the convergence problem has been solved for arbitrary $\alpha, \beta, \gamma, \delta$ in Horton [1]), and the Cotes' numbers are positive when $\gamma = \frac{1}{2}\alpha + \frac{3}{4}$ if the following inequality holds:

$$(9.23) \qquad \sum_{k=0}^{n} \frac{(\alpha + \frac{1}{2})_k (\frac{1}{2}\alpha + \frac{5}{4})_k (\frac{1}{2})_k}{(\alpha + 1)_k (\frac{1}{2}\alpha + \frac{1}{4})_k k!} \frac{P_k^{(\alpha, -1/2)}(x)}{P_k^{(-1/2, \alpha)}(1)} > 0,$$

$$-1 < x \leq 1, \quad n = 0, 1, \cdots.$$

When $\alpha > \frac{1}{2}$ this would follow from

$$\sum_{k=0}^{n} \frac{P_k^{(\alpha, -1/2)}(x)}{P_k^{(-1/2, \alpha)}(1)} \geq 0, \qquad\qquad -1 \leq x \leq 1,$$

which unfortunately has only been proved when $\alpha = \frac{1}{2}, \frac{3}{2}, \frac{5}{2}$. However, (9.23) is true when $\alpha \geq \frac{1}{2}$. Gasper has recently proved this. So I have changed my mind about the problem of proving the positivity of Cotes' numbers. It is more than just a test problem, it is the source of new and interesting inequalities. Since sources of inequalities are rare it is clear that further work should be done on this problem.

To see that Vietoris' inequalities are deep consider the following problem. Let

$$q(\theta) = \sum_{k=1}^{n} \lambda_{n-k} \sin k\theta.$$

If $\lambda_0 > \lambda_1 \geq \lambda_2 \geq \cdots \geq \lambda_{n-1} \geq 0$, then $q(\theta)$ has $n - 1$ zeros which satisfy

$$(9.24) \qquad k\pi/(n + \tfrac{1}{2}) < \theta_k < (k + 1)\pi/(n + \tfrac{1}{2}), \quad k = 1, 2, \cdots, n - 1.$$

Pólya [1] proved the zeros were real and Szegö [5] gave the estimates. If the additional inequalities

$$(9.25) \quad 2\lambda_0 - \lambda_1 > \lambda_1 - \lambda_2 \geq \lambda_2 - \lambda_3 \geq \cdots \geq \lambda_{n-2} - \lambda_{n-1} \geq \lambda_{n-1} \geq 0$$

are satisfied, then the further inequalities

$$(9.26) \qquad\qquad\qquad \theta_k < (k + \tfrac{1}{2})\pi/n, \qquad\qquad k = 1, 2, \cdots, n,$$

hold (see Szegö [5]). The inequalities (9.24) show that the zeros are uniformly distributed, and the inequalities (9.24) and (9.26) show that they are uniformly separated. Vietoris' inequalities can be used to show that

$$(9.27) \qquad\qquad\qquad (2k - 1)\lambda_{k-1} \geq 2k\lambda_k > 0, \qquad\qquad k = 1, 2, \cdots,$$

imply

$$(9.28) \qquad k\pi/(n + \tfrac{1}{4}) < \theta_k < (k + \tfrac{1}{2})\pi/(n + \tfrac{1}{4}), \quad k = 1, 2, \cdots, n - 1,$$

(see Askey–Steinig [1]). The conditions (9.25) imply

$$(9.29) \qquad\qquad\qquad \lambda_k \geq Ak, \quad A > 0, \quad k = 1, 2, \cdots,$$

while the conditions (9.27) only imply

(9.30) $\lambda_k \geqq Ak^{1/2}, \quad A > 0, \quad k = 1, 2, \cdots .$

Neither of the conditions (9.25) or (9.27) implies the other, but the inequalities (9.29) and (9.30) show that (9.27) is in some sense weaker than (9.25), while the conclusions (9.28) are slightly (but only very slightly) stronger than those in (9.24) and (9.26). This gives an indication of the depth of Vietoris' inequalities.

To return to the inequality (9.2), one of the most surprising developments is in a recent paper of Gasper [8]. As was remarked above, Makai [1] proved (9.2) when $\lambda = 0$, and the other extreme case $\lambda = \alpha - \frac{1}{2}$ was proved by Fields and Ismail [1] by an incredible asymptotic argument. Since the two methods used by Makai and Fields–Ismail had nothing in common, Gasper decided to try to find other proofs of these two results and he succeeded in writing both of these integrals as sums of squares of Bessel functions times a product of gamma functions, and these gamma functions were clearly positive when $\alpha > \frac{1}{2}$. In the general case, $0 < \lambda < \alpha - \frac{1}{2}$ he obtained the coefficients as a $_5F_4$ with a free parameter, which he chose to reduce the coefficient to a Saalschützian $_4F_3$, which was then transformed by Whipple's formula to a well-poised $_7F_6$ which was clearly positive, since all the terms in the sum were positive.

To wax philosophical for a minute, I would like to suggest that Hardy [1, p. 14] was wrong when he suggested that the age of great formulas may be over. Whipple's formula was found shortly after Ramanujan died, and it is clearly a very important formula. So is the addition formula of Šapiro and Koornwinder. Great formulas do not come along very often, but they never have. The rate at which important formulas are being found is probably decreasing, especially as a proportion of the mathematics being done, but there are still important formulas to be found. And when an explicit formula can be found there is nothing to beat it. In the rush to abstraction and generalization we often forget this.

Gasper [8] has a number of other interesting positive integrals, including

(9.31)
$$\int_0^x (x^2 - t^2)^\lambda t^{\lambda + 1/2} J_\alpha(t) \, dt \geqq 0,$$

$$0 \leqq \lambda \leqq \alpha - \tfrac{1}{2}, \quad x \geqq 0.$$

It would be interesting to consider

(9.32)
$$\int_0^x (x^\mu - t^\mu)^\lambda t^{\lambda + 1/2} J_\alpha(t) \, dt$$

for $0 < \mu < 2$. The next easiest case is probably $\mu = \frac{2}{3}$.

Additional sums and integrals could be considered, but we shall forgo this here, and go on to a few applications, and then close with a brief comment about the general problem which has really been considered in these two lectures.

One application is to univalent functions. Let

(9.33)
$$f(z) = z + \sum_{n=2}^\infty a_n z^n.$$

Define

$$\Delta a_n = a_n - a_{n+1},$$

$$\Delta^{k+1} a_n = \Delta(\Delta^k a_n).$$

Fejér [8] proved that $f(z)$ is univalent for $|z| < 1$ if $\Delta^j a_n \geqq 0, j = 0, 1, 2, 3, 4, n = 0,$
$1, \cdots$. His proof used

$$\sum_{k=0}^{n} \frac{(4)_{n-k}}{(n-k)!}(k+1)\sin(k+1)\theta > 0, \qquad 0 < \theta < \pi,$$

which is (1.19). Szegö [7] improved this by reducing the conditions to $\Delta^j a_n \geqq 0,$
$j = 0, 1, 2, 3$. He used

$$\sum_{k=0}^{n} \frac{(3)_{n-k}}{(n-k)!}(k+1)\sin(k+1)\theta > 0,$$

$$0 < \theta \leqq \gamma, \quad \frac{\pi}{2} < \gamma \leqq \frac{2\pi}{3}.$$

When $\gamma = 2\pi/3$ this is Schweitzer's inequality (1.24). Robertson [2] and Fuchs [2]
gave a refinement to p-valent functions, and each of their proofs used the inequality

(9.34) $$\sum_{k=0}^{n} \frac{(3)_{n-k}}{(n-k)!}(k+1)\sin(k+\tfrac{1}{2})\theta > 0, \qquad 0 < \theta < \pi.$$

They each proved (9.34) without stating it in this form, and so neither of them
realized that they could have used Fejér's inequality

$$\sum_{k=0}^{n} \frac{(3)_{n-k}}{(n-k)!}(k+\tfrac{1}{2})\sin(k+\tfrac{1}{2})\theta > 0, \qquad 0 < \theta < \pi.$$

This is (1.21). For (1.1) is

$$\sum_{k=0}^{n} \sin(k+\tfrac{1}{2})\theta \geqq 0,$$

and then

$$\sum_{k=0}^{n} \frac{(3)_{n-k}}{(n-k)!}\sin(k+\tfrac{1}{2})\theta \geqq 0,$$

so (9.34) trivially follows from (1.21).

Fuchs [1] has shown that no difference less than the third will give univalence for
$f(z)$ even when fractional differences are used. This is related to the fact that (1.21)
is best possible in the sense that

$$\sum_{k=0}^{n} \frac{(3-\varepsilon)_{n-k}}{(n-k)!}(k+\tfrac{1}{2})\sin(k+\tfrac{1}{2})\theta$$

changes sign in any interval $0 \leqq \theta \leqq \theta_0, \theta_0 > 0$, for infinitely many n when $\varepsilon > 0$.

These inequalities can be used to construct positive definite functions, i.e., functions which are Fourier transforms of positive measures. If $f(x)$ is even, continuous for $x \geq 0$, vanishes at infinity and convex for $x > 0$, then the Fourier transform of f is nonnegative and so $f(x)$ is positive definite by the inversion formula (see Pólya [1], [3]). One analogue of even functions in R^n is the class of radial functions. Let

$$f(x_1, \cdots, x_n) = g((x_1^2 + \cdots + x_n^2)^{1/2}).$$

If $g(t)$ is continuous for $t \geq 0$, $(-1)^j g^{(j)}(t) \geq 0, j = 0, 1, \cdots, [n/2], t > 0, \lim_{t \to \infty} g(t) = 0$, and $(-1)^{[n/2]} g^{[n/2]}(t)$ is convex for $t > 0$, then $f(x_1, \cdots, x_n)$ is positive definite (see Askey [17]). An analogous theorem for spheres and projective spaces is in Askey [11]. The first theorem of this type was found by W. H. Young [2]. There are also theorems of this type which give sufficient conditions for a function to be the Fourier transform of a unimodal distribution (see Askey [15]). Hopefully other applications will arise after these inequalities become better known.

The real question which has been considered in these two lectures is to find real intervals over which generalized hypergeometric functions do not change sign, or on which they are positive. There are related questions on sets in the complex plane which are also interesting, and only a start has been made on the deep results in the plane. Szegö [2] proved

(9.35)
$$\sum_{k=0}^{n} P_k(x) z^k \neq 0, \qquad -1 \leq x \leq 1, \quad |z| < 1,$$

and a result of J. W. Alexander [1] can be used to prove

(9.36)
$$\sum_{k=0}^{n} \frac{\sin (k + 1)\theta}{(k + 1) \sin \theta} z^k \neq 0, \qquad 0 \leq \theta \leq \pi, \quad |z| < 1.$$

For this and further results, see Bustoz [1].

Added in proof. The first application of Theorem 9.1 to comparing areas under arches seems to have been given by Hartman and Wintner [1]. L. Lorch called this paper to my attention.

LECTURE 10

Suggestions for Further Work

A number of open problems and conjectures have been given in the previous lectures. But only a small part of mathematics has been covered, and problems involving special functions arise in many other areas. I am not aware of all the current applications of special functions, and am surprised almost monthly by new applications. For example, Krawtchouk polynomials and the dual Hahn polynomials (they are called Eberlein polynomials by Delsarte) are useful in coding theory (see Delsarte [1]). This is not too surprising, since coding theory deals with discrete problems, so it is natural to suspect that difference equations will be used, and hypergeometric functions satisfy some of the few difference equations which can be solved. They also arise in population genetics for a similar reason (see Karlin–McGregor [3]).

Ginibre [1] proved that

$$(10.1) \qquad \int_{-\infty}^{\infty} \int_{-\infty}^{\infty} \prod_{i=1}^{N} [H_{n_i}(x) \pm H_{n_i}(y)] e^{-x^2 - y^2} \, dx \, dy \geq 0,$$

where $H_n(x)$ is the Hermite polynomial of degree n and all possible choices of plus and minus signs are taken. He also proved a similar inequality for $P_n^{(-1/2, -1/2)}(x)$, where the integration is with respect to $(1 - x^2)^{-1/2}(1 - y^2)^{-1/2} \, dx \, dy$ over $-1 \leq x, y \leq 1$. A similar inequality should be proved for $P_n^{(\alpha, \alpha)}(x)$, $\alpha > -\frac{1}{2}$, $(-1)^n L_n^\alpha(x)$, $\alpha > -1$ and $K_n(x, p, N)$, $\frac{1}{2} \leq p < 1$.

Grünbaum [1] proved that

$$(10.2) \qquad\qquad P_n(x) + 1 \geq P_r(x) + P_s(x), \qquad\qquad 0 \leq x \leq 1,$$

when $n(n + 1) = r(r + 1) + s(s + 1)$, and then used this to prove

$$(10.3) \qquad\qquad 1 + J_0(z) \geq J_0(x) + J_0(y), \quad z^2 = x^2 + y^2$$

(see Grünbaum [2]). A direct proof is in Askey [9], as well as the inequality

$$(10.4) \qquad 1 + \mathcal{J}_\alpha(z) \geq \mathcal{J}_\alpha(x) + \mathcal{J}_\alpha(y), \quad z^2 = x^2 + y^2, \quad \alpha \geq 0,$$

where

$$(10.5) \qquad\qquad \mathcal{J}_\alpha(x) = \Gamma(\alpha + 1) 2^\alpha x^{-\alpha} J_\alpha(x),$$

so

$$\mathcal{J}_\alpha(0) = 1.$$

It would be interesting to find the smallest α for which (10.4) holds.

Nelson [1] obtain a fascinating property of

$$K_c(x, y) = \sum_{n=0}^{\infty} \frac{c^n H_n(x) H_n(y)}{2^n n!}.$$

If

$$\int_{-\infty}^{\infty} |f(x)|^p e^{-x^2} dx$$

is finite and

$$f(x) \sim \sum_{n=0}^{\infty} a_n H_n(x),$$

$$\Gamma(c) f(x) \sim \sum_{n=0}^{\infty} c^n a_n H_n(x),$$

then

$$\left[\frac{1}{2\sqrt{\pi}} \int_{-\infty}^{\infty} |\Gamma(c) f(x)|^p e^{-x^2} dx \right]^{1/p} \leq \left[\frac{1}{2\sqrt{\pi}} \int_{-\infty}^{\infty} |f(x)|^q e^{-x^2} \right]^{1/q}$$

when $1 \leq q \leq p \leq \infty, 0 \leq c \leq \sqrt{(q-1)/(p-1)}$, and when $c > \sqrt{(q-1)/(p-1)}$ the operator is unbounded. So it is either a contraction or it is unbounded.

It is likely that similar results hold for Laguerre, Meixner, and Charlier expansions.

Each of the above inequalities was suggested by a problem in physics. There are times when physicists are decades behind mathematicians (as well as the other way around). For example, Racah [1] rediscovered Dixon's 1903 formula for the well-poised $_3F_2$, and Regge [1] rediscovered one of the Thomae transformation formulas from 1879 (see Huszár [1]). Similar symmetries for the $6 - j$ symbol (or Saalschützian $_4F_3$) were found by Whipple [2], [3] years before physicists needed them. So it is clear that people who know something about hypergeometric functions should cultivate at least one physicist friend with whom they can exchange information. Also the Journal of Mathematical Physics and Communications on Mathematical Physics should be looked at regularly.

Physicists are better than most other people in the notation they use, and they are usually (but not always) aware when they are considering a function which has been useful enough in the past to rate a name. But this is not true of most people who work on combinatorial problems. Sums of products of binomial coefficients are not quite everywhere dense in combinatorial problems, but they occur with sufficient regularity to make it worthwhile keeping track of what happens there. People who work in this area desperately need to be told that Euler, Gauss, Kummer, Riemann, Saalschütz, Dixon, and Dougall worked on these problems, and that it is not necessary to derive every formula from the Chu–Vandermonde sum each time it appears. Recently these sums have appeared in group theory (see Miech [1], [2], but I have not had time to see what role hypergeometric functions can play with respect to his problems). Binomial coefficients play an important role in the analysis of computer algorithms (see Knuth [1]). This book, and the other ones in this set are very impressive, but the comments on binomial coefficients are only partially correct. There are actually very few different sums of products of binomial coefficients which can be summed, and each new one which has a number of free parameters is very important. For a fashionable and modern view of the role of special functions in computer science, see Moses [1]. I feel he has a totally incorrect view of this subject. As these lectures have tried to demonstrate, Laplace, Euler, Gauss, and many others knew what they were doing when they studied certain special classes of differential equations in detail. It is impossible to say very much about a subject which is very general. Solutions to an arbitrary differential equation will have very few properties and it is only when the differential equation has some structure that a rich theory can be developed. For reasons which are only beginning to be understood, the differential equations of the special functions which were discovered in the eighteenth and nineteenth centuries are exactly the differential equations with sufficient structure so that really interesting facts can be discovered about their solutions. And these are also the functions and differential equations which come up in many branches of science and mathematics. So it makes a lot of sense to treat them as something special and to analyze them in detail, rather than to try only to derive properties of them which also hold for solutions to much more general differential equations. This would restrict our knowledge far too much. It seems very reasonable to try to automate the general program that Miller has been developing using Lie algebras to study special functions. But it seems far too restricted to try to do this in a much more general setting, for we know that most of the interesting formulas for the classical functions do not extend beyond them, at least in usable form.

Because the classical functions, and the series that represent them, arise in many ways, there are many ways to look at them, and so it is not surprising that there are many different ways of evaluating those few sums which can be evaluated. This seems to be the answer to a question which has bothered me for years. Cauchy's theorem is very powerful and it is clearly one of the best ways of evaluating sums. Yet when the most important sums are considered, almost none of them was discovered by use of Cauchy's theorem. A plausible explanation for this is that most of these sums were evaluated by very few people and by and large these people used methods they developed when they were quite young. Since these sums are quite fundamental, they can be evaluated by many methods and so elementary (but often quite complicated) methods will work. People who build systems are not very good at summing series, I know because I am not one of the fortunate few who is good at summing series and so have to rely on a system to aid me.

Another good source of problems is univalent function theory. For example, Wilken [1] proved

$$(10.6) \quad |{}_2F_1(-n, -q; q; z)| \leqq {}_2F_1(-n, -q; q; 1) = \frac{(2q)_n}{(q)_n}, \qquad q \geqq \tfrac{1}{2}, \quad |z - \tfrac{1}{2}| \leqq \tfrac{1}{2},$$

Brannan–Clunie–Kirwan [1] proved

$$|_2F_1(-n, -\alpha; -n - \alpha + 1; z)| \leq |_2F_1(-n, -\alpha; -n - \alpha + 1; -1)|,$$
(10.7)
$$\alpha \geq 2, \quad |z| \leq 1,$$

and conjectured that it held for $\alpha \geq 1$, and Aharonov–Friedlund [1] proved the stronger inequality

(10.8) $$|_2F_1(-n, 1; -n + \alpha + 1; e^{i\theta})| \leq |_2F_1(-n, 1; -n + \alpha + 1; -1)|, \quad \theta \text{ real,}$$

$\alpha \geq 1$. None of these inequalities was stated in terms of hypergeometric functions and hypergeometric functions were not used in the proofs. Wall [1] proved the inequality

$$\left| _2F_1(\alpha, 1; \gamma; z) - \frac{(\gamma - 1)^2}{(\gamma - 1)^2 - \alpha^2} \right| \leq \frac{\alpha(\gamma - 1)}{(\gamma - 1)^2 - \alpha^2},$$
(10.9)
$$|z| \leq 1, \quad 0 < \alpha < \gamma - 1,$$

and he used continued fractions in this proof. It is not unreasonable to suppose that almost two hundred years worth of work on hypergeometric functions would have developed methods of treating these inequalities. And if such methods have not been developed, they should be. There are now enough special results to give some idea of what type of general result to look for.

Problems will often arise in a way so that the hypergeometric character of the problem is not immediately obvious. For example, last winter I received a letter from I. Joo asking if I could prove

$$\frac{d^{2n}}{dx^{2n}}[(1 - x)^{-\alpha}(1 + x)^{-\beta}] > 0, \quad -1 < x < 1, \quad \alpha, \beta \geq 0.$$

I could, because I could translate the problem into a problem about $_2F_1$'s and could use a result of Stieltjes (see Szegö [9]), about the zeros of such functions.

I would strongly recommend looking carefully at Karlin–Szegö [1]. There are still many interesting problems tied up with higher order determinants of the classical polynomials.

When Szegö wrote his book [9] he could say that the most interesting results of the recent past were on general orthogonal polynomials. In the last ten years it is clear that the most important work on orthogonal polynomials was done on the classical polynomials in one variable. This will almost surely not be true in the next ten years, but it is not clear if orthogonal polynomials in several variables will be the place to look for the really interesting problems (see Koornwinder [6], [8] for a start), if there will be a new breakthrough on general orthogonal polynomials, or if the fairly special polynomials like those of Pollaczek will be the source of the most interesting results. I tend to feel it will be either the first or the third. Szegö's theory of general orthogonal polynomials was a great achievement and clearly the last word has not been said about it. But it is a theory which generalizes the classical polynomials, and it has been pushed far enough so that surprising new results are unlikely. Many of the convergence theorems are already so strong that they cannot be improved. So to get a further generalization we must look to some nonclassical orthogonal polynomials and see what properties they have. Clearly the Pollaczek polynomials are the most interesting nonclassical polynomials in one variable which have been found up to the present, and it is likely that the next breakthrough for general orthogonal polynomials in one variable will come only after we understand Pollaczek polynomials more thoroughly.

Another source of interesting problems is a computing machine (used intelligently). The graphs of the zeros of the partial sums of e^z (see Iverson [1] and Newman–Rivlin [1]) suggest a number of problems, and not all of them were solved by Newman and Rivlin. Other hypergeometric functions should be examined to see what is really happening here. Bessel polynomials are not very interesting orthogonal polynomials, but they seem to be interesting hypergeometric functions, and the known results about their zeros need to be extended to other hypergeometric functions (see Wimp [1]).

Added in proof. A beautiful proof and extension of Grünbaum's inequality (10.3) was found by Mercer [2]. He used an integral representation of Bessel functions which he had proved earlier (see Mercer [1]).

Beckner [1], who was motivated by some ideas in Nelson [1], proved the strong form of the Hausdorff–Young inequality for Fourier transforms. This very important result deals with the series (2.44W) when r is purely imaginary, while Nelson's theorem deals with real r. Clearly this problem should be considered for general complex r in the unit circle. Analogous problems need to be considered for other classical orthogonal polynomials.

Saff and Varga [1] have found a very nice proof of the existence of zero free parabolic regions for the partial sums of the exponential function and many other functions.

Another application of hypergeometric functions in physics, this time Legendre functions, is contained in Cornille and Martin [1], [2].

References

ABRAMOWITZ, M. AND I. STEGUN

[1] *Handbook of Mathematical Functions*, N.B.S. Applied Math. Ser. 55 (1964), Washington, D.C.

ADAMS, J. C.

[1] *On the expression of the product of any two Legendre's coefficients by means of a series of Legendre's coefficients*, Proc. Royal Soc. London, 27 (1878), pp. 63–71.

AHARONOV, D. AND S. FRIEDLAND

[1] *On an inequality connected with the coefficient conjecture for functions of bounded boundary rotation*, Ann. Acad. Sci. Fenn. Ser. AI, 524 (1972).

ALEXANDER, J. W., II

[1] *Functions which map the interior of the unit circle upon simple regions*, Ann. of Math., 17 (1915), pp. 12–22.

ASKEY, R.

[1] *Orthogonal expansions with positive coefficients*, Proc. Amer. Math. Soc., 16 (1965), pp. 1191–1194.

[2] *Mehler's integral for $P_n(\cos \theta)$*, Amer. Math. Monthly, 76 (1969), pp. 1046–1049.

[3] *Orthogonal polynomials and positivity*, Studies in Applied Mathematics 6, Wave Propagation and Special Functions, D. Ludwig and F. W. J. Olver, eds., SIAM, Philadelphia, 1970, pp. 64–85.

[4] *Linearization of the product of orthogonal polynomials*, Problems in Analysis, R. Gunning, ed., Princeton University Press, Princeton, N.J., 1970, pp. 223–228.

[5] *Orthogonal expansions with positive coefficients. II*, SIAM J. Math. Anal., 2 (1971), pp. 340–346.

[6] *Positivity of the Cotes numbers for some Jacobi abscissas*, Numer. Math., 19 (1972), pp. 46–48.

[7] *Positive Jacobi polynomial sums*, Tôhoku Math. J., 2nd ser., 24 (1972), pp. 109–119.

[8] *Mean convergence of orthogonal series and Lagrange interpolation*, Acta Math. Acad. Sci. Hungar., 23 (1972), pp. 71–85.

[9] *Grünbaum's inequality for Bessel functions*, J. Math. Anal. Appl., 41 (1973), pp. 122–124.

[10] *Summability of Jacobi series*, Trans. Amer. Math. Soc., 179 (1973), pp. 71–84.

[11] *Refinement of Abel summability for Jacobi series*, Proc. Symp. Pure Math. vol. 26, Harmonic Analysis on Homogeneous Spaces, C. Moore, ed., American Mathematical Society, Providence, R.I., 1973, pp. 335–338.

[12] *Certain rational functions whose power series have positive coefficients. II*, SIAM J. Math. Anal., 5 (1974), pp. 53–57.

[13] *Jacobi polynomials, I, New proofs of Koornwinder's Laplace type integral representation and Bateman's bilinear sum*, Ibid., 5 (1974), pp. 119–124.

[14] *Some absolutely monotonic functions*, Studia Sci. Math., 9 (1974), pp. 51–56.

[15] *Some characteristic functions of unimodal distributions*, J. Math. Anal. Appl., 50 (1975), to appear.

[16] *Positive Jacobi polynomial sums, III, Linear Operators and Approximations, II*, P. L. Butzer and B. Sz. Nagy, eds., Birkhäuser Verlag Basel, 1975.

[17] *Radial characteristic functions*, Mathematics Research Center, Tech. Rep. 1262, University of Wisconsin, Madison.

[18] *A note on the history of series*, Mathematics Research Center Tech. Rep. 1532, University of Wisconsin, Madison.

ASKEY, R. AND J. FITCH

[1] *Positivity of the Cotes numbers for some ultraspherical abscissas*, SIAM J. Numer. Anal., 5 (1968), pp. 199–201.

[2] *Solution to Problem 67-6, A Trigonometric inequality, by J. N. Lyness and C. B. Moler*, SIAM Rev., 11 (1969), pp. 82–86.

[3] *Integral representations for Jacobi polynomials and some applications*, J. Math. Anal. Appl., 26 (1969), pp. 411–437.

[4] *A positive Cesàro mean*, Publ. Elek. Fak. Univ. Beogradu, ser. Mat. i Fiz., 406 (1972), pp. 119–122.

ASKEY, R., J. FITCH AND G. GASPER

[1] *On a positive trigonometric sum*, Proc. Amer. Math. Soc., 19 (1968), p. 1507.

ASKEY, R. AND G. GASPER

[1] *Linearization of the product of Jacobi polynomials, III*, Canad. J. Math., 23 (1971), pp. 332–338.

[2] *Jacobi polynomial expansions of Jacobi polynomials with non-negative coefficients*, Proc. Camb. Phil. Soc., 70 (1971), pp. 243–255.

[3] *Certain rational functions whose power series have positive coefficients*, Amer. Math. Monthly, 79 (1972), pp. 327–341.

[4] *Convolution structures for Laguerre polynomials*, J. Analyse Math., to appear.

[5] *Positive Jacobi polynomial sums, II*, Amer. J. Math, to appear.

ASKEY, R., G. GASPER AND M. ISMAIL

[1] *A positive sum from summability theory*, J. Approximation Theory, 13 (1975), to appear.

ASKEY, R. AND H. POLLARD

[1] *Some absolutely monotonic and completely monotonic functions*, SIAM J. Math. Anal., 5 (1974), pp. 58–63.

ASKEY, R. AND J. STEINIG

[1] *Some positive trigonometric sums*, Trans. Amer. Math. Soc., 187 (1974), pp. 295–307.

[2] *A monotonic trigonometric sum*, Amer. J. Math., to appear.

ASKEY, R. AND S. WAINGER

[1] *A transplantation theorem for ultraspherical coefficients*, Pacific J. Math., 16 (1966), pp. 393–405.

[2] *A dual convolution structure for Jacobi polynomials*, Proc. Conference on Orthogonal Expansions and their Continuous Analogues, D. Haimo, ed., Southern Illinois University Press, Carbondale, 1967, pp. 25–36.

ATKINSON, F. V.

[1] *Discrete and Continuous Boundary Problems*, Academic Press, New York, 1964.

BAILEY, W. N.

[1] *On the product of two Legendre polynomials*, Proc. Camb. Phil. Soc., 29 (1933), pp. 173–177.

[2] *Generalized Hypergeometric Series*, Cambridge University Press, Cambridge, 1935.

[3] *The generating function of Jacobi polynomials*, J. London Math. Soc., 13 (1938), pp. 8–11.

[4] *Contiguous hypergeometric functions of the type $_3F_2(1)$*, Proc. Glasgow Math. Assoc., 2 (1954), pp. 62–65.

BATEMAN, H.

[1] *A generalization of the Legendre polynomial*, Proc. London Math. Soc. (2), 3 (1905), pp. 111–123.

[2] *The solution of linear differential equations by means of definite integrals*, Trans. Camb. Phil. Soc., 21 (1909), pp. 171–196.

[3] *Partial Differential Equations of Mathematical Physics*, Cambridge University Press, Cambridge, 1932.

BECKNER, W.

[1] *Inequalities in Fourier analysis on R^n*, Proc. Nat. Acad. Sci. U.S.A., 61 (1975), to appear.

BINGHAM, N. H.

[1] *Integral representations for ultraspherical polynomials*, J. London Math. Soc. (2), 6 (1973), pp. 1–11.

BOAS, R. P., JR.

[1] *Entire Functions*, Academic Press, New York, 1953.

BOCHNER, S.

[1] *Hilbert distances and positive definite functions*, Ann. of Math. (2), 42 (1941), pp. 647–656.

[2] *Positive zonal functions on spheres*, Proc. Nat. Acad. Sci. U.S.A., 40 (1954), pp. 1141–1147.

BRAAKSMA, B. L. J. AND B. MEULENBELD

[1] *Jacobi polynomials as spherical harmonics*, Indag. Math., 30 (1968), pp. 384–389.

BRANNAN, D. A., J. G. CLUNIE AND W. E. KIRWAN

[1] *On the coefficient problem for functions of bounded boundary rotation*, Ann. Acad. Sci. Fenn. Ser. AI, 523 (1973).

BROWN, J. W. AND J. L. GOLDBERG

[1] *Generalized Appell connection sequences*, J. Math. Anal. Appl., 46 (1974), pp. 242–248.

BURCHNALL, J. L. AND A. LAKIN

[1] *The theorems of Saalschütz and Dougall*, Quart. J. Math. Oxford (2), 1 (1950), pp. 161–164.

BUSTOZ, J.

[1] *Jacobi polynomial sums and univalent Cesàro means*, Proc. Amer. Math. Soc., to appear.

CARLITZ, L.

[1] *On a problem in the history of Chinese mathematics*, Mat. Lapok., 6 (1955), pp. 219–220.

[2] *A formula of Bateman*, Proc. Glasgow Math. Assoc., 3 (1957), pp. 99–101.

[3] *Some generating functions for the Jacobi polynomials*, Bull. U.M.I. (3), 16 (1961), pp. 150–155.

[4] *The product of two ultraspherical polynomials*, Proc. Glasgow Math. Assoc., 5 (1961–2), pp. 76–79.

CARLSON, B. C.

[1] *New proof of the addition formula for Gegenbauer polynomials*, SIAM J. Math. Anal., 2 (1971), pp. 347–351.

CARTAN, E.

[1] *Sur la détermination d'un système orthogonal complet dans un espace de Riemann symétrique clos*, Rend. Circ. Mat. Palermo, 53 (1929), pp. 217–252.

CHÉBLI, H.

[1] *Sur la positivité des opérateurs de "translation généralisée" associés à un opérateur de Sturm-Liouville sur* $[0, \infty[$. C.R. Acad. Sci. Paris, Ser A, 275 (1972), pp. 601–604.

CHOLEWINSKI, F. M. AND D. T. HAIMO

[1] *Classical analysis and the generalized heat equation*, SIAM Rev., 10 (1968), pp. 67–80.

CHRYSTAL, G.

[1] *Algebra*, vol. 2, 2nd ed., A. and C. Black, London, 1900.

CHU, SHIH-CHIEH

[1] *Ssu Yuan Yü Chien (Precious Mirror of the Four Elements)*, 1303 (Chinese).

CHURCH, R. F.

[1] *On a constant in the theory of trigonometric series*, Math. Comp., 19 (1965), p. 501.

CLAUSEN, T.

[1] *Ueber die Fälle, wenn die Reihe von der Form*

$$y = 1 + \frac{\alpha}{1} \cdot \frac{\beta}{\gamma} x + \frac{\alpha \cdot \alpha + 1}{1 \cdot 2} \cdot \frac{\beta \cdot \beta + 1}{\gamma \cdot \gamma + 1} x^2 + \text{etc.}$$

ein Quadrat von der Form

$$z = 1 + \frac{\alpha'}{1} \cdot \frac{\beta'}{\gamma'} \cdot \frac{\delta'}{\varepsilon'} x + \frac{\alpha' \cdot \alpha' + 1}{1 \cdot 2} \cdot \frac{\beta' \cdot \beta' + 1}{\gamma' \cdot \gamma' + 1} \cdot \frac{\delta' \cdot \delta' + 1}{\varepsilon' \cdot \varepsilon' + 1} x^2 + \text{etc.}$$

hat., J. Reine Angew Math., 3 (1828), pp. 89–91.

COIFMAN, R. R. AND G. WEISS

[1] *Representations of compact groups and spherical harmonics*, L'Ens. Math., 14 (1968), pp. 121–173.

COPSON, E. T.

[1] *An Introduction to the Theory of Functions of a Complex Variable*, Oxford University Press, Oxford, 1935.

CORNILLE, H. AND A. MARTIN

[1] *Constraints on the phase of scattering amplitudes due to positivity*, Nuclear Physics, B49 (1972), pp. 413–440.

[2] *Constraints on the phases of helicity amplitudes due to positivity*, Ibid., B77 (1974), pp. 141–162.

DAVIS, J. AND I. I. HIRSCHMAN, JR.

[1] *Toeplitz forms and ultraspherical polynomials*, Pacific J. Math., 18 (1966), pp. 73–95.

DELSARTE, P.

[1] *An algebraic approach to the association schemes of coding theory*, Philips Res. Reports Suppl., 10 (1973).

DELSARTE, P., J. M. GOETHALS AND J. J. SEIDEL

[1] *Bounds for systems of lines, and Jacobi polynomials*, to appear.

DINGHAS, A.

[1] *Zur Darstellung einiger Klassen hypergeometrischer Polynome durch Integrale von Dirichlet-Mehlerschen Typus*, Math. Zeit., 53 (1950), pp. 76–83.

DOUGALL, J.

[1] *On Vandermonde's theorem and some more general expansions*, Proc. Edinburgh Math. Soc., 25 (1907), pp. 114–132.

[2] *A theorem of Sonine in Bessel functions, with two extensions to spherical harmonics*, Ibid., 37 (1919), pp. 33–47.

[3] *The product of two Legendre polynomials*, Proc. Glasgow Math. Assoc., 1 (1952/3), pp. 121–125.

DUNKL, C. F.

[1] *An expansion in ultraspherical polynomials with nonnegative coefficients*, SIAM J. Math. Anal., 5 (1974), pp. 51–52.

[2] *A Krawtchouk polynomial addition theorem and wreath products of symmetric groups*, to appear.

DUNKL, C. F. AND D. E. RAMIREZ

[1] *Krawtchouk polynomials and the symmetrization of hypergroups*, Ibid., 5 (1974), pp. 351–366.

DURAND, L.

[1] *Nicholson-type integrals for products of Gegenbauer functions*, Abstract in Notices Amer. Math. Soc., 20 (1973), pp. 704–B24.

EAGLESON, G. K.

[1] *A characterization theorem for positive definite sequences on the Krawtchouk polynomials*, Austral. J. Statist., 11 (1969), pp. 29–38.

ERDÉLYI, A.

[1] *Transformation of hypergeometric integrals by means of fractional integration by parts*, Quart. J. Math. (Oxford), 10 (1939), pp. 176–189.

ERDÉLYI, A., ET AL.

[1] *Higher Transcendental Functions*, vol. 1, McGraw-Hill, New York, 1953.

[2] *Higher Transcendental Functions*, vol. 2, McGraw-Hill, New York, 1953.

EULER, L.

[1] *Specimen transformationis singularis serierum*, Nova acta academiae scientiarum Petropolitanae, 12 (1794), 1801, pp. 58–70; reproduced in Opera Omnia, 16 (part 2) (1935), pp. 41–55.

FAVARD, J.

[1] *Sur les polynômes de Tchebycheff*, C. R. Acad. Sci. Paris, 200 (1935), pp. 2052–2055.

FEJÉR, L.

[1] *Sur le développement d'une fonction arbitraire suivant les fonctions de Laplace*, C. R. Acad. Sci. Paris, 146 (1908), pp. 224–225; reproduced in Gesammelte Arbeiten, I, pp. 319–322.

[2] *Über die Laplacesche Reihe*, Math. Ann., 67 (1909), pp. 76–109; reproduced in Gesammelte Arbeiten, I, pp. 503–537.

[3] *Einige Sätze, die sich auf das Vorzeichen, u.s.w.*, Monatsh. Math. Phys., 35 (1928), pp. 305–344; reproduced in Gesammelte Arbeiten, II, pp. 202–237.

[4] *Ultraspàrikus polynomok összegéröl*, Matés Fiz. Lapok, 38 (1931), pp. 161–164; *Über die Summe ulträspharischer Polynome*, reproduced in Gesammelte Arbeiten, II, pp. 421–423.

[5] *Mechanische Quadraturen mit positiven Cotesschen Zahlen*, Math. Zeit., 37 (1933), pp. 289–309; reproduced in Gesammelte Arbeiten, II, pp. 457–478.

[6] *Gestaltliches über die Partialsummen und ihre Mittelwerte bei der Fourrierreihe und der Potenzreihe*, Zeit. Angew. Math. Mech., 13 (1933), pp. 80–88; reproduced in Gesammelte Arbeiten, II, pp. 479–492.

[7] *Neue Eigenschaften der Mittlewerte bei den Fourierreihen*, J. London Math. Soc., 8 (1933), pp. 53–62; reproduced in Gesammelte Arbeiten, II, pp. 493–501.

[8] *Trigonometrische Reihen und Potenzreihen mit mehrfach monotoner Koeffizientenfolge*, Trans. Amer. Math. Soc., 39 (1936), pp. 18–59; reproduced in Gesammelte Arbeiten, II, pp. 581–620.

FELDHEIM, E.

[1] *On the positivity of certain sums of ultraspherical polynomials*, J. Analyse Math., 11 (1963), pp. 275–284.

[2] *Contribution à la theorie des polynomes de Jacobi*, Mat. Fiz. Lapok, 48 (1941), pp. 453–504 (Hungarian, French summary).

FELLER, W.

[1] *Infinitely divisible distributions and Bessel functions associated with random walks*, SIAM J. Appl. Math., 14 (1966), pp. 864–875.

FERRERS, N. M.

[1] *An Elementary Treatise on Spherical Harmonics and Subjects Connected with Them*, Macmillan, London, 1877.

FIELDS, J.

[1] *Asymptotic expansions of a class of hypergeometric polynomials with respect to the order, III*, J. Math. Anal. Appl., 12 (1965), pp. 593–601.

FIELDS, J. AND M. ISMAIL

[1] *On the positivity of some $_1F_2$'s*, SIAM J. Math. Anal., 6 (1975), pp. 551–559.

FLENSTED-JENSEN, M.

[1] *Paley-Wiener type theorems for a differential operator connected with symmetric spaces*, Ark. Mat., 10 (1972), pp. 143–162.

[2] *The spherical functions on the universal covering of $SU(n-1, 1)/SU(n-1)$*, Mat. Inst. Kbhvn. Univ. Preprint Series, 1 (1973).

FLENSTED-JENSEN, M. AND T. H. KOORNWINDER

[1] *The convolution structure for Jacobi function expansions*, Ark. Mat., 11 (1973), pp. 145–162.

FOLLAND, G. B.

[1] *Spherical harmonic expansion of the Poisson-Szegö kernel for the ball*, Proc. Amer. Math. Soc., to appear.

FREUD, G.

[1] *Orthogonale Polynome*, Birkhäuser-Verlag, Basel and Stuttgart, 1969.

FUCHS, I.

[1] *Potenzreihen mit mehrfach monotonen Koeffizienten*, Arch. Math., 22 (1971), pp. 275–278.

[2] *Power series with multiply monotonic coefficients*, Math. Ann., 190 (1971), pp. 289–292.

GANGOLLI, R.

[1] *Positive definite kernels on homogeneous spaces and certain stochastic processes related to Lévy's Brownian motion of several parameters*, Ann. Inst. H. Poincaré, Sect. B, 3 (1967), pp. 121–226.

GASPER, G.

[1] *Nonnegative sums of cosine, ultraspherical and Jacobi polynomials*, J. Math. Anal. Appl., 26 (1969), pp. 60–68.

[2] *Linearization of the product of Jacobi polynomials. I*, Canad. J. Math., 22 (1970), pp. 171–175.

[3] *Linearization of the product of Jacobi polynomials. II*, Ibid., 22 (1970), pp. 582–593.

[4] *Positivity and the convolution structure for Jacobi series*, Ann. of Math., 93 (1971), pp. 112–118.

[5] *Banach algebras for Jacobi series and positivity of a kernel*, Ibid., 95 (1972), pp. 261–280.

[6] *Nonnegativity of a discrete Poisson kernel for the Hahn polynomials*, J. Math. Anal. Appl., 42 (1973), pp. 438–451.

[7] *Projection formulas for orthogonal polynomials of a discrete variable*, Ibid., 45 (1974), pp. 176–198.

[8] *Positive integrals of Bessel functions*, SIAM J. Math. Anal., 6 (1975), to appear.

[9] *Formulas of the Dirichlet-Mehler type*, to appear.

104   REFERENCES

GEGENBAUER, L.
  [1] *Zur Theorie der Functionen X_n^m*, Sitz. Akad. Wiss. Wien, Math.-Naturw. Kl., 66 (2), (1872),
      pp. 55–62.
  [2] *Uber einige bestimmte Integrale*, Sitz. Math. Natur. Klasse Akad. Wiss. Wien, 70 (2), (1875),
      pp. 433–443.
  [3] *Zur Theorie der Functionen $C_n^\nu(x)$*, Denkschriften der Akademie der Wiss. in Wien, Math.
      Naturwiss. Kl., 48 (1884), pp. 293–316.
GILBERT, R. P.
  [1] *Function Theoretic Methods in Partial Differential Equations*, Academic Press, New York, 1969.
GILLIS, J. AND G. WEISS
  [1] *Products of Laguerre polynomials*, M.T.A.C. (now Math. Comp.), 14 (1960), pp. 60–63.
GINIBRE, J.
  [1] *General formulation of Griffiths' inequalities*, Comm. Math. Phys., 16 (1970), pp. 310–328.
GLASSER, M. L.
  [1] *Some definite integrals of the product of two Bessel functions of the second kind: (order zero)*,
      Math. Comp., 28 (1974), pp. 613–615.
GRONWALL, T. H.
  [1] *Über die Gibbssche Erscheinung und die trigonometrischen Summen* $\sin x + 1/2 \sin 2x + \cdots$
      $+(1/n) \sin nx$, Math. Ann., 72 (1912), pp. 228–243.
GRÜNBAUM, F.
  [1] *A property of Legendre polynomials*, Proc. Nat. Acad. Sci., U.S.A., 67 (1970), pp. 959–960.
  [2] *A new kind of inequality for Bessel functions*, J. Math. Anal. Appl., 41 (1973), pp. 115–121.
HARDY, G. H.
  [1] *Ramanujan*, Cambridge University Press, Cambridge, 1940.
  [2] *Further researches in the theory of divergent series and integrals*, Trans. Camb. Phil. Soc.,
      21 (1908), pp. 1–48, reprinted in Collected Papers of G. H. Hardy, vol. VI, 1974, pp. 214–262.
HARTMAN, P.
  [1] *On differential equations and the function $J_\mu^2 + Y_\mu^2$*, Amer. J. Math., 83 (1961), pp. 154–188.
  [2] *On differential equations, Volterra equations and the function $J_\mu^2 + Y_\mu^2$*, Ibid., 95 (1973), pp.
      553–593.
HARTMAN, P. AND G. S. WATSON
  [1] *"Normal" distribution functions on spheres and the modified Bessel functions*, Annals of Prob.,
      2 (1974), pp. 593–607.
HARTMAN, P. AND A. WINTNER
  [1] *On nonconservative linear oscillators of low frequency*, Amer. J. Math., 70 (1948), pp. 529–539.
HELGASON, S.
  [1] *The Radon transform on Euclidean spaces, compact two-point homogeneous spaces and Grassman
      manifolds*, Acta Math., 113 (1965), pp. 153–180.
HENRICI, P.
  [1] *Addition theorems for general Legendre and Gegenbauer functions*, J. Rational Mech. Anal.
      (now Indiana J. Math.), 4 (1955), pp. 983–1018.
  [2] *My favorite proof of Mehler's integral*, Amer. Math. Monthly, 78 (1971), pp. 183–185.
HERMITE, C. AND T. J. STIELTJES
  [1] *Correspondence d'Hermite et de Stieltjes*, vol. 2, Gauthier-Villars, Paris, 1905, p. 43.
HILLE, E.
  [1] *Note on some hypergeometric series of higher order*, Proc. London Math. Soc., 4 (1929),
      pp. 50–54.
HIRSCHMAN, I. I., JR.
  [1] *Variation diminishing Hankel transforms*, J. Analyse Math., 8 (1960–61), pp. 307–336.
  [2] *Extreme eigenvalues of Toeplitz forms associated with ultraspherical polynomials*, J. Math.
      Mech. (now Indiana J. Math.), 13 (1964), pp. 249–282.
  [3] *Eigenvalues of Toeplitz operators on SU(2)*, Duke Math. J., 41 (1974), pp. 51–82.
HOBSON, E. W.
  [1] *The Theory of Spherical and Ellipsoidal Harmonics*, Cambridge University Press, Cambridge,
      1931.

HOLLÓ, Á.

  [1] *A mechanikus quadraturáról*, Thesis, Budapest, 1939, 23p.

HORTON, R. L.

  [1] *Expansions using orthogonal polynomials*, Ph.D. thesis, University of Wisconsin, Madison, 1973.

  [2] *Jacobi polynomials, IV, A family of variation diminishing kernels*, SIAM J. Math. Anal., 6 (1975), pp. 544–550.

HSÜ, H. Y.

  [1] *Certain integrals and infinite series involving ultraspherical polynomials and Bessel functions*, Duke Math. J., 4 (1938), pp. 374–383.

HUA, L. K.

  [1] *Harmonic Analysis of Functions of Several Complex Variables in the Classical Domains*, Trans. Math. Monographs 6, 1963, American Mathematical Society, Providence, R.I.

HUSZÁR, M.

  [1] *Symmetries of Wigner coefficients and Thomae–Whipple functions*, Acta Phys. Acad. Sci. Hungar., 32 (1972), pp. 181–185.

HYLLERAAS, E.

  [1] *Linearization of products of Jacobi polynomials*, Math. Scand., 10 (1962), pp. 189–200.

HYLTÉN-CAVALLIUS, C.

  [1] *A positive trigonometrical kernel*, Tolfte Skand. Mat. Kongr. 1953 Lund (1954), pp. 90–94.

IGARI, S. AND Y. UNO

  [1] *Banach algebras related to the Jacobi polynomials*, Tôhoku Math. J., 21 (1969), pp. 668–673.

IVERSON, K.

  [1] *The zeros of the partial sums of e^z*, M.T.A.C. (now Math. Comp.), 7 (1953), pp. 165–168.

JACKSON, D.

  [1] *Über eine trigonometrische Summe*, Rend. Circ. Mat. Palermo, 32 (1911), pp. 257–262.

JACOBI, C. G. J.

  [1] *Untersuchungen über die Differentialgleichung der hypergeometrischen Reihe*, J. Reine Angew. Math., 56 (1859), pp. 149–165. Gesammelte Werke, Vol. 6, pp. 184–202.

KALUZA, T.

  [1] *Elementarer Beweis einer Vermutung von K. Friedrichs und H. Lewy*, Math. Zeit., 37 (1933), pp. 689–697.

KARLIN, S. AND J. MCGREGOR

  [1] *The differential equations of birth-and-death processes, and the Stieltjes moment problem*, Trans. Amer. Math. Soc., 85 (1957), pp. 489–546.

  [2] *Classical diffusion processes and total positivity*, J. Math. Anal. Appl., 1 (1960), pp. 163–183.

  [3] *The Hahn polynomials, formulas and an application*, Scripta Math., 26 (1961), pp. 33–46.

KARLIN, S. AND G. SZEGÖ

  [1] *On certain determinants whose elements are orthogonal polynomials*, J. Analyse Math., 8 (1961), pp. 1–157.

KNUTH, D.

  [1] *The Art of Computer Programming, Vol. 1, Fundamental Algorithms*, 2nd ed., Addison-Wesley, Reading, Mass., 1973.

KOGBETLIANTZ, E.

  [1] *Recherches sur la sommabilité des séries ultraspériques par la méthode des moyennes arithmetiques*, J. Math. Pures Appl. (9), 3 (1924), pp. 107–187.

KOORNWINDER, T. H.

  [1] *The addition formula for Jacobi polynomials. I, Summary of results*, Indag. Math., 34 (1972), pp. 188–191.

  [2] *The addition formula for Jacobi polynomials. II, The Laplace type integral representation and the product formula*, Math. Centrum Amsterdam, Rep. TW 133 (1972).

  [3] *The addition formula for Jacobi polynomials. III, Completion of the proof*, Math. Centrum Amsterdam, Rep. TW 135 (1972).

  [4] *The addition formula for Jacobi polynomials and spherical harmonics*, SIAM J. Appl. Math., 25 (1973), pp. 236–246.

[5] *Jacobi polynomials. II, An analytic proof of the product formula*, SIAM J. Math. Anal., 5 (1974), pp. 125–137.

[6] *Orthogonal polynomials in two variables which are eigenfunctions of two algebraically independent partial differential operators. I, II*, Indag. Math., 36 (1974), pp. 48–58; pp. 59–66.

[7] *Jacobi polynomials, III, An analytic proof of the addition formula*, SIAM J. Math. Anal., 6 (1975), pp. 533–543.

[8] *Jacobi polynomials and their two-variable analogues*, Thesis, University of Amsterdam, 1974.

[9] *A new proof of a Paley-Wiener type theorem for the Jacobi transform*, Math. Centrum Amsterdam, Rep. TW 143 (1974).

[10] *The addition formula for Jacobi polynomials and the theory of orthogonal polynomials in two variables, a survey*, Math. Centrum Amsterdam, Rep. TW 145 (1974).

KOSCHMIEDER, L. AND R. STROMAN

[1] *Zwei Lösungen der Aufgabe 77*, Jahr. Deutsch. Math. Verein., 43 (1933), pp. 64–66.

KOSHLIAKOV, N. S.

[1] *On Sonine's polynomials*, Mess. Math., 55 (1926), pp. 152–160.

KUMMER, E. E.

[1] *Über die hypergeometrische Reihe*, J. Reine Angew. Math., 15 (1836), pp. 39–83, 127–172.

LAPLACE, P.

[1] *Théorie des attractions des sphéroides et de la figure des planètes*, Mem. de l'Acad. Royale des Sciences de Paris (1782) (published in 1785), pp. 113–196; reprinted in Oeuvres de Laplace, 10 (1894), pp. 339–419.

LORCH, L.

[1] *Comparison of two formulations of Sonine's theorem and of their respective applications to Bessel functions*, Studia Sci. Math. Hungar., 1 (1966), pp. 141–145.

LORCH, L., M. E. MULDOON AND P. SZEGO

[1] *Higher monotonicity properties of certain Sturm-Liouville functions, IV*, Canad. J. Math., 24 (1972), pp. 349–368.

LORENTZ, G. G. AND K. ZELLER

[1] *Abschnittslimitierbarkeit und der Satz von Hardy-Bohr*, Arch. Math. (Basel), 15 (1964), pp. 208–213.

LUKE, Y. L., W. FAIR, G. COOMBS and R. MORAN

[1] *On a constant in the theory of trigonometric series*, Math. Comp., 19 (1965), pp. 501–502.

MACMAHON, P. A.

[1] *Combinatory Analysis*, vol. 1, Cambridge University Press, Cambridge, 1915.

MAKAI, E.

[1] *On a monotonic property of certain Sturm-Liouville functions*, Acta Math. Acad. Sci. Hungar., 3 (1952), pp. 165–171.

[2] *An integral inequality satisfied by Bessel functions*, Ibid., 25 (1974), pp. 387–390.

MANOCHA, H. L.

[1] *Some formulae involving Appell's function F_4*, Publ. Inst. Math. (Beograd) (N.S.), 9 (23), (1969), pp. 153–156.

MARX, A.

[1] *Aufgaben 77*, Jahr. Deutsch. Math. Verein., 39 (1930), p. 1.

MEHLER, F. G.

[1] *Über eine mit den Kugelfunktionen und Cylinder funktionen verwandte Funktion und ihre Anwendung in der Theorie der Elektrizitätsverteilung*, Math. Ann., 18 (1881), pp. 161–194.

MERCER, A. McD.

[1] *On certain functional identities in E^N*, Canad. J. Math., 23 (1971), pp. 315–324.

[2] *Grunbaüm's inequality for Bessel functions and extensions of it*, SIAM J. Math. Anal., to appear.

MIECH, R. J.

[1] *Some p groups of maximal class*, Trans. Amer. Math. Soc., 189 (1974), pp. 1–47.

[2] *Counting commutators*, Ibid., 189 (1974), pp. 49–61.

MILLER, W.

[1] *Lie Theory and Special Functions*, Academic Press, New York, 1968.

[2] *Special functions and the complex Euclidean group in 3-space, II*, J. Math. Phys., 9 (1968), pp. 1175–1187.

MOSES, J.

[1] *Towards a general theory of special functions*, Comm. A.C.M., 15 (1972), pp. 550–554.

MÜLLER, C.

[1] *Spherical Harmonics*, Lecture Notes in Mathematics, no. 17, Springer-Verlag, Berlin, 1966.

MULLIN, R. AND G. C. ROTA

[1] *On the foundations of combinatorial theory, III: Theory of binomial enumeration*, Graph Theory and its Applications, B. Harris, ed., Academic Press, New York, 1970, pp. 167–213.

NEEDHAM, J.

[1] *Science and Civilization in China, vol. 3, Mathematics and the Sciences of the Heavens and the Earth*, Cambridge University Press, New York, 1959.

NELSON, E.

[1] *The free Markoff field*, J. Functional Anal., 12 (1973), pp. 211–227.

NEUMANN, F. E.

[1] *Beiträge zur Theorie der Kugelfunctionen*, Leipzig, 1878.

VON NEUMANN, J. AND I. J. SCHOENBERG

[1] *Fourier integrals and metric geometry*, Trans. Amer. Math. Soc., 50 (1941), pp. 226–251.

NEWMAN, D. J. AND T. J. RIVLIN

[1] *The zeros of the partial sums of the exponential function*, J. Approximation Theory, 5 (1972), pp. 405–412.

NIELSON, N.

[1] *Théorie des fonctions métasphériques*, Paris, 1911.

NOVIKOFF, A.

[1] *On a special system of orthogonal polynomials*, Dissertation, Stanford University, 1954, available from University Microfilms, Ann Arbor.

ORIHARA, A.

[1] *Hermitian polynomials and infinite-dimensional motion group*, J. Math. Kyoto Univ., 6 (1966), pp. 1–12.

PARRISH, C.

[1] *Multivariate umbral calculus*, Ph.D. thesis, University of California at San Diego, La Jolla, 1974.

PEETRE, J.

[1] *The Weyl transform and Laguerre polynomials*, Matematiche, 27 (1972), pp. 301–323.

PFAFF, J. F.

[1] *Disquisitiones Analyticae*, Helmstadii, 1797.

[2] *Observationes analyticae ad L. Euleri Institutiones Calculi Integralis*, vol. IV, Supplem. II et IV, Historie de 1793, Nova acta academiae scientiarum Petropolitanae, Tom XI, 1797, pp. 38–57. (Note, the history section is paged separately from the scientific section of this journal.)

PÓLYA, G.

[1] *Über die Nullstellen gewisser ganzer Funktionen*, Math. Zeit., 2 (1918), pp. 352–383.

[2] *Über die Konvergenz von Quadraturverfahren*, Ibid., 37 (1933), pp. 264–286.

[3] *Remark on characteristic functions*, Proc. First Berkeley Symp. on Stat. and Prob., University of California Press, Berkeley, 1949, pp. 63–78.

PÓLYA, G. AND I. J. SCHOENBERG

[1] *Remarks on the de la Vallée Poussin means and convex conformal maps of the circle*, Pacific J. Math., 8 (1958), pp. 295–334.

RACAH, G.

[1] *Theory of complex spectra, II*, Phys. Rev., 62 (1942), pp. 438–462.

RAINVILLE, E. R.

[1] *Special Functions*, Macmillan, New York, 1960.

RANKIN, R. A.

[1] *Functions whose powers have non-negative Taylor coefficients*, Proc. London Math. Soc., 14A (1965), pp. 239–248.

REGGE, T.

[1] *Symmetry properties of Clebsch-Gordan's coefficients*, Nuovo Cimento (10), 10 (1958), pp. 544–545.

ROBERTSON, M. S.

[1] *The coefficients of univalent functions*, Bull. Amer. Math. Soc., 51 (1945), pp. 733–738.

[2] *Power series with multiply monotonic coefficients*, Mich. Math. J., 16 (1969), pp. 27–31.

ROBIN, L.

[1] *Fonctions sphériques de Legendre et fonctions sphéroidales*, vol. 3, Gauthier-Villars, Paris, 1959.

ROGOSINSKI, W. AND G. SZEGÖ

[1] *Über die Abschnitte von Potenzreihen, die in einem Kreise beschränkt bleiben*, Math. Zeit., 28 (1928), pp. 73–94.

ROTA, G.-C., D. KAHANER AND A. ODLYZKO

[1] *On the foundations of combinatorial theory, VIII*, J. Math. Anal. Appl., 42 (1973), pp. 684–760.

RUDIN, W.

[1] *Fourier Analysis on Groups*, Interscience, New York, 1962.

SAFF, E. AND R. VARGA

[1] *Zero-free parabolic regions for sequences of polynomials*, SIAM J. Math. Anal., to appear.

ŠAPIRO, R. L.

[1] *Special functions related to representations of the group $SU(n)$, of class I with respect to $SU(n - 1)(n \geqq 3)$*, Izv. Vyssh. Uchebn. Zaved. Matematika (1968), no. 4 (71), pp. 97–107. (In Russian.)

SARMANOV, I. O.

[1] *A generalized symmetric gamma correlation*, Dokl. Akad. Nauk SSSR, 179 (1968), pp. 1276–1278; Soviet Math. Dokl., 9 (1968), pp. 547–550.

SARMANOV, O. V. AND Z. N. BRATOEVA

[1] *Probabilistic properties of bilinear expansions of Hermite polynomials*, Theory Prob. Applications, 12 (1967), pp. 470–481.

SCHINDLER, S.

[1] *Some transplantation theorems for the generalized Mehler transform and related asymptotic expansions*, Trans. Amer. Math. Soc., 155 (1971), pp. 257–291.

SCHOENBERG, I. J.

[1] *Metric spaces and completely monotonic functions*, Ann. of Math., (2), 39 (1938), pp. 811–841.

[2] *Positive definite functions on spheres*, Duke Math. J., 9 (1942), pp. 96–108.

SCHWEITZER, M.

[1] *The partial sums of second order of the geometric series*, Ibid., 18 (1951), pp. 527–533.

SEIDEL, W. AND O. SZÁSZ

[1] *On positive harmonic functions and ultraspherical polynomials*, J. London Math. Soc., 26 (1951), pp. 36–41.

SONINE, N. J.

[1] *Recherches sur les fonctions cylindriques et le développement des fonctions continues en séries*, Math. Ann., 16 (1880), pp. 1–80.

STEINIG, J.

[1] *The sign of Lommel's function*, Trans. Amer. Math. Soc., 163 (1972), pp. 123–129.

[2] *A criterion for the positivity of sine polynomials*, Proc. Amer. Math. Soc., 38 (1973), pp. 583–586.

STIRLING, J.

[1] *Methodus differentialis; sive, Tractatus de summatione et interpolatione serierum infinitarum*, London, 1730.

Szegö, G.

[1] *Koeffizientenabschätzungen bei ebenen und räumlichen harmonischen Entwicklungen*, Math. Ann., 96 (1927), pp. 601–632.

[2] *Zur Theorie der Legendreschen Polynome*, Jahr. Deutsch. Math. Verein., 40 (1931), pp. 163–166.

[3] *Asymptotische Entwicklungen der Jacobischen Polynome*, Schr. der König. Gelehr. Gesell. Naturwiss. Kl., 10 (1933), pp. 33–112.

[4] *Über gewisse Potenzreihen mit lauter positiven Koeffizienten*, Math. Zeit., 37 (1933), pp. 674–688.

[5] *Inequalities for the zeros of Legendre polynomials and related functions*, Trans. Amer. Math. Soc., 39 (1936), pp. 1–17.

[6] *On some Hermitian forms associated with two given curves of the complex plane*, Ibid., 40 (1936), pp. 450–461.

[7] *Power series with multiply monotonic sequences of coefficients*, Duke Math. J., 8 (1941), pp. 559–564.

[8] *On the relative extrema of Legendre polynomials*, Boll. Un. Mat. Ital., (3), 5 (1950), pp. 120–121.

[9] *Orthogonal Polynomials*, Colloquium Publications, vol. 23, 3rd ed., American Mathematical Society, Providence, R.I., 1967.

Takács, L.

[1] *On an identity of Shih-Chieh Chu*, Acta Sci. Math. (Szeged), 34 (1973), pp. 383–391.

Thomae, J.

[1] *Über die Funktionen welche durch Reihen von der Form dargestellt werden:* $1 + pp'p''/1q'q'' + \cdots$, J. Reine Angew. Math., 87 (1879), pp. 26–73.

Titchmarsh, E. C.

[1] *Some integrals involving Hermite polynomials*, J. London Math. Soc., 23 (1948), pp. 15–16.

Turán, P.

[1] *On a trigonometrical sum*, Ann. Soc. Polonaise Math., 25 (1952), pp. 155–161.

[2] *On some problems in the theory of the mechanical quadrature*, Mathematica (Cluj), (31), 8 (1966), pp. 181–192.

Tyan, S.-G.

[1] *The structure of bivariate distribution functions and their relation to Markov processes*, Ph.D. thesis, Princeton University, 1975.

Tyan, S. and J. B. Thomas

[1] *Characterization of a class of bivariate distribution functions*, J. Multivariate Analysis, (1975), to appear.

de la Vallée Poussin, C. J.

[1] *Sur l'approximation des fonctions de variables réelles et de leurs dérivées par des polynomes et des suites limitées de Fourier*, Bull. de l'Acad. Roy. de Belgique (Classe des Sciences), no. 3, 1908.

Vandermonde, A.

[1] *Mémoire sur des irrationnelles de différens ordres avec une application au cercle*, Mem. Acad. Roy. Sci. Paris (1772), pp. 489–498.

Vere-Jones, D.

[1] *Finite bivariate distributions and semigroups of nonnegative matrices*, Quart. J. Math., Oxford (2), 22 (1971), pp. 247–270.

Vietoris, L.

[1] *Über das Vorzeichen gewisser trigonometrischer Summen*, Sitzungsber. Oest. Akad. Wiss., 167 (1958), pp. 125–135, and Anzeiger Oest. Akad. Wiss. (1959), pp. 192–193.

Vilenkin, N. Ja.

[1] *Some relations for Gegenbauer functions*, Uspekhi Matem. Nauk (N.S.), 13 (1958), no. 3 (81), pp. 167–172. (In Russian.)

[2] *Special Functions and the Theory of Group Representations*, Translations of Math. Monographs, vol. 22, American Mathematical Society, Providence, 1968.

Wall, H. S.

[1] *A class of functions bounded in the unit circle*, Duke Math. J., 7 (1940), pp. 146–153.

WATSON, G. N.

[1] *Another note on Laguerre polynomials*, J. London Math. Soc., 14 (1939), pp. 19–22.

[2] *A Treatise on the Theory of Bessel Functions*, 2nd ed., Cambridge University Press, Cambridge, 1944.

[3] *A reduction formula*, Proc. Glasgow Math. Assoc., 2 (1954), pp. 57–61.

WEINBERGER, H.

[1] *A maximum property of Cauchy's problem*, Ann. of Math. (2), 64 (1956), pp. 505–513.

WHIPPLE, F. J. W.

[1] *A group of generalized hypergeometric series: relations between 120 allied series of the type* $F[^{a,b,c}_{e,f}]$, Proc. London Math. Soc.'(2), 23 (1924), pp. 104–114.

[2] *Well-poised series, generalized hypergeometric series having parameters in pairs, each pair having the same sum*, Ibid., 24 (1925), pp. 247–263.

[3] *Well-poised series and other generalized hypergeometric series*, Ibid., 25 (1926), pp. 525–544.

WHITTAKER, E. T. AND G. N. WATSON

[1] *A Course of Modern Analysis*, 4th ed., Cambridge University Press, Cambridge, 1952.

WILKEN, D. R.

[1] *The integral means of close-to-convex functions*, Mich. Math. J., 19 (1972), pp. 377–379.

WILSON, M. W.

[1] *On the Hahn polynomials*, SIAM J. Math. Anal., 1 (1970), pp. 131–139.

[2] *Nonnegative expansions of polynomials*, Proc. Amer. Math. Soc., 24 (1970), pp. 100–102.

[3] *On a new discrete analogue of the Legendre polynomials*, SIAM J. Math. Anal., 3 (1972), pp. 157–169.

WIMP, J.

[1] *On the zeros of a confluent hypergeometric function*, Proc. Amer. Math. Soc., 16 (1965), pp. 281–283.

YOUNG, W. H.

[1] *On a certain series of Fourier*, Proc. London Math. Soc. (2), 11 (1912), pp. 357–366.

[2] *On the Fourier series of bounded functions*, Ibid., 12 (1913), pp. 41–70.

ZAREMBA, S. K.

[1] *Some properties of polynomials orthogonal over the set* $\langle 1, 2, \cdots, N \rangle$, Mathematics Research Center, Tech. Rep. 1342, University of Wisconsin, Madison.

ZERNIKE, F. AND H. C. BRINKMAN

[1] *Hypersphärische Funktionen und die in sphärischen Bereichen orthogonalen Polynome*, Nederl. Akad. Wetensch. Proc., 38 (1935), pp. 161–170.